有機樂園

Grow Organic @home

小家庭的種植攻略

嘉道理農場暨植物園 編著

萬里機構・萬里書店

有機樂園：小家庭的種植攻略	Grow Organic @ home
編著及翻譯 嘉道理農場暨植物園	Author and Translators Kadoorie Farm and Botanic Garden
編輯 鄧宇雁	Editor Rain Tang
攝影 嘉道理農場暨植物園、黃德生	Photographers Kadoorie Farm and Botanic Garden, Wong Tak Sang
封面設計 朱靜	Cover Design Ching Chu
版面設計 辛紅梅　劉葉青	Layout Design Cindy Xin, Rosemary Liu

出版者
萬里機構 • 萬里書店
香港鰂魚涌英皇道1065號
東達中心1305室
電話
傳真
電郵
網址

Publisher
Wan Li Book Company
Room 1305, Eastern Centre, 1065 King's Road,
Quarry Bay, Hong Kong.
Tel:　　2564 7511
Fax:　　2565 5539
Email: info@wanlibk.com
Web Site:　　http://www.wanlibk.com
　　　　　　http://www.facebook.com/wanlibk

發行者
香港聯合書刊物流有限公司
香港新界大埔汀麗路36號
中華商務印刷大廈3字樓
電話
傳真
電郵

Distributor
SUP Publishing Logistics (HK) Ltd.
3/F., C&C Building, 36 Ting Lai Road,
Tai Po, N.T., Hong Kong
Tel:　　2150 2100
Fax:　　2407 3062
Email: info@suplogistics.com.hk

承印者
美雅印刷製本有限公司

Printer
Elegance Printing & Book Binding Co Ltd.

出版日期
二零一六年一月第一次印刷
二零一七年八月第二次印刷

Publishing Date
First print in January 2016
Second print in August 2017

萬里機構　萬里 Facebook

本書以再造紙及大豆油墨印刷
This book is printed with recycled paper and soy ink

引言

現今，香港超過百分之九十的食物，都是從約一百五十個國家進口。全球性的食物系統，極度地依賴長程運輸、工業化運作，以及大規模地使用化學肥料及農藥。運輸食物所耗用的化石能源，還遠高於食物本身的能量，涉及的溫室氣體排放量佔全球三分之一。

而支持着這個不合常理系統的，就是廉價石油。因此，我們食物的價格，與油價的關係密不可分。

我們必須清楚知道，為超過七百萬人口提供穩定糧食，是一個沉重的議題。在石油頂峰（全球石油產量逐步下降，廉價石油日子將會結束）、氣候轉變及人口過多的影響下，所面臨的挑戰也日益艱巨。

人們逐漸認識到，我們的城市正面對這些難題，越來越多人關注到環境、社會及經濟的可持續性。這些，促使世界各地的市政府及公民社會，擴闊思維，思考如何透過永續都市農業的方法，加強社區的生存及適應能力。

都市農業，泛指於城內或城郊種植植物和飼養動物。由於香港城市化的程度甚高，香港境內的所有農業活動（包括較遙遠新界地區），均可歸類為都市農業。

都市農業最顯著的特徵，是它與城市生態融為一體。這意味着都市農業能夠從城市的人口中吸納勞動力，同時亦有機會獲得城市的資源（例如以有機廢物作為堆肥材料、採用經淨化的污水作灌溉）。因此，都市農業與城市消費者有着直接的聯繫，並對城市生態有直接的影響。都市農業是城市食物系統重要的一環，亦為野生動物提供了棲息地、保持生態系統的韌力、與其他城市設施共享土地，它在城市的政策和規劃上應當佔一席位。都市農業絕非過去的遺俗，也非農村移民帶來市區的產物。它是城市不可或缺的一部分，應當鼓勵發展。

發展都市農業，需要更多公眾對本土食物生產的支持及參與。栽種食物有時並非想像般複雜。誠然，我們面對空間不足及環境方面的種種限制，但許多情況下，我們仍能夠在家居中種植一些食物。

我們衷心希望，在這個全球面臨嚴峻危機的時刻，這本書能讓更多的人，成為都市農業運動的一份子。

嘉道理農場暨植物園

Introduction

Today, over 90% of the food consumed in Hong Kong is imported from around 150 countries. The globalised food system relies heavily on long-distance transport, industrialized operations, and extensive input of synthetic fertilizers and pesticides. The fossil fuel energy consumed in moving food far exceeds the food energy provided by the food and accounts for 1/3 of today's global greenhouse gas emissions.

The price of our food is closely linked to the price of oil and this illogical system is fueled only by the availability of cheap oil. It is therefore important to realise that 'food security' for our population of over seven million is a very serious issue and will be more challenging under the escalating effects of Peak Oil (the gradual reduction in global oil production and the end of cheap oil), climate change and over-population.

While the public awareness of the vulnerability of cities to these challenges is growing, there are also ever-increasing issues of concern related to environmental, social and economic sustainability, which drive leading municipal governments and civic-societies worldwide to think more creatively about how to greatly increase community resilience and livability, in relation to food and urban sustainable agriculture.

Urban Agriculture may be defined, briefly, as the growing of plants and the raising of animals within and around cities. Given that Hong Kong is highly urbanized, all local agricultural activities within our boundaries, even the remote New Territories, can be categorised as Urban Agriculture.

The most striking feature of urban agriculture, distinguishing it from rural agriculture, is that it is embedded in the urban ecosystem. This means that manpower is drawn from urban residents; there is potential access to urban resources (like organic waste for compost, and urban grey water for irrigation), there is direct linkage with urban consumers and direct impact on urban ecology. Urban Agriculture is part of the urban food system, provides habitats for wildlife – maintaining ecological resilience – shares land-use with other urban functions, and is subject to urban policies and town planning. Urban agriculture is not a relic of the past that will fade away, nor is it something brought to the city by rural immigrants. It is an integral part of the urban system that should be encouraged to increase and should involve everyone.

Expanding urban agriculture requires a much wider public participation and public support for local food production. Growing some of one's own food is a lot less complicated than people might think. Shortage of space and environmental constraints can be challenging, but it is usually possible to grow some food at home.

We sincerely hope that this book creates an entry point for more people to take part in the urban agriculture movement, in a time of severe global crisis.

Kadoorie Farm and Botanic Garden

目錄

Chapter 01

社區 農圃與家居 種植
Community Farming
and Growing Food at Home

農夫以種田為生，然而，「生產」並不是種植的唯一價值。今時今日，社區農圃已遍及世界各地的農地、市區公園、學校、社區中心、公共空間及屋邨，發揮多元化的作用，例如提供輔助食物渠道、教育、休閑康樂、提高社區凝聚力、促助有機資源循環及綠化等等。雖然「城市農夫」的實際種植面積，較正式農場小得多，但它為無數人提供栽種食物的體驗，連結泥土、心靈與社會。

由於作用不同，農圃之間的設計與運作形式亦會有差異，以下有一些例子：

Farmers live by farming. Nevertheless, 'production' is not the sole value of farming. Today, community farms are set up worldwide on farmland, urban parks, schools, community centres, public space and housing estates to cope with diverse community needs such as providing food supplements, education, recreation, leisure, community cohesion, organic resource recovery, greening and other purposes. Although the actual size of farmland subscribed by each 'city farmer' is far smaller than a typical farm, community farms offer opportunity for a much larger number of people to experience food growing, together. It reconnects soil, soul and society.

The design and operational format of community farms varies according to their core function. Here are some examples:

社區農圃
Community Farming

康樂

康樂及休閒是這類農圃的主要目的。它們座落於鄉郊、住宅及其他公眾地方，參加的「城市農夫」源自社會中不同的階層，都是追求自然與農耕體驗的人。部份參加者將它當成團結家庭的聚會，也有些人會作為社交及休閒的活動。

農圃內，農地會細分為一個個小區，分配給各個參加單位。與一般鄉郊或市區公園比較，社區農圃可以讓參加者有更多參與及互動空間，他們可以一同學習、一同種植、施肥，以及分享成果。

Recreation

Recreation and leisure are the primary purposes of community farms. Community farms are now being set up on rural farmland, in housing estates and public spaces to serve a broad range of users, the so-called 'holiday-farmers', who look for a recreational experience related to nature and farming. Some users see it as a means to tighten family-bonding, while some see it as social and leisure activities.

At a community farm, farmland is sub-divided into small plots for allocating to subscribers. Compared to Country or urban parks, the users' experience at a community farm is far more participatory and interactive - they share tools and ideas, learn together, plant, fertilise and harvest their crops, all at the same farm.

大埔舊墟公立學校(寶湖道)與區內組織合作，讓小孩與長者一同合作興建園圃和耕種，促進跨代聯繫，達至長幼共融。

Tai Po Old Market Public School (Plover Cove) collaborates with local NGOs to enable cross-generational integration, between primary students and elderly people through community farming.

教育

主要由學校或環保團體舉辦，教育是這類形農圃的最主要目標。教育性農圃一般面積較小，產量亦並非其最優先考慮；最重要的，是透過農耕的活動，讓參加者得到接觸泥土、接觸自然的學習機會。參加者亦需經常留意農作物的生長情況，以及組織相關的農務工作。

Education

Education is often the objective of farms that are set up by schools or environmental organisations. The size of these farms are usually small. Rather than yields, organisers are more interested in the learning opportunities that the farms can offer participants as they come into contact with soil and nature. Participants are often engaged to observe plant-growth and perform farming duties in an organized way.

有機耕種已廣泛融入環境教育活動之中。
Organic farming is widely used in environmental education programmes.

食物

現在，許多城市人（包括香港）都已認真審視都市種植所帶來的廣泛裨益。而提供食物當然是其中一個主要目標，越來越多人加入這個行列，以創意的方法，在露台、天台等居住空間種植美味而健康的食物。

Food

More city dwellers around the world, including Hong Kong, are taking a serious look at urban farming for its wide range of benefits. Among which, growing some food is one of the key purposes. Workshops and finding creative ways for converting living spaces of balconies, rooftops and the indoor environment for delicious and healthy food production use are getting more popular.

農圃雖小，生產力卻一點也不低。
Although small in size, this community farm is highly productive.

社區凝聚力

社區農圃可促進跨代溝通及共融，增進自我的幸福價值，這些優點已得到地區研究的確認。

這些農圃往往由宗教、慈善團體，又或是非牟利組織所舉辦，目的是增加凝聚力；也有一些組織會以種植來實踐其社會使命，如復康工作。

Social cohesion and mission

The merits of community farming in improving inter-generational cooperation and communication, as well as subjective well-being, have been confirmed by local research. Community farms have been set up by religious groups, charities, and non-profit making organizations for social cohesion or clearly defined social missions, such as rehabilitation.

在康樂園，管理公司將廚餘循環與有機種植活動連結一起，成功帶動住戶有興趣參與和投入。
At Hong Lok Yuen, the management office integrates organic food growing activities with its food waste recovery programme, which successfully arouses residents' interest and long-term commitment.

美化

在都市園景設計中加入農作物元素，以提供新功能、增加多樣性及生產力的做法，已經日益普遍。不少地區的餐廳、酒店及學院，亦以「食物花園」作為公眾的吸引點。

Aesthetics

There is a growing interest in using crops in landscape design to bring in new functions and to enrich the diversity and productivity of urban landscapes. Edible landscaping is also getting more widely adopted in local restaurants, hotels and institutions as a public attraction.

食用蔬果也能具備觀賞價值。
An edible landscape can be aesthetically pleasing.

農圃設計
Farm Design

農圃性質 Primary Purpose	設計原則 Design Principles
食物生產 Food production	✤ 儘量擴大種植面積,包括利用垂直空間 ✤ 設立育苗區、堆肥箱及其他促助生產的設施 ✤ 以能夠持續性提供食物為設計目標 ✤ Maximise farming area, including vertical space, for food production. ✤ Reserve space for placing a seedling nursery, compost bin and other items that are essential for enhancing outputs. ✤ Able to supply food continuously throughout the year.
教育 Education	✤ 選擇在教育方面具代表性的農作物 ✤ 從教育角度,考慮農圃的空間及通道設計 ✤ 在顯眼位置設立指示牌 ✤ 預留空間存放農具、文具及其他活動所需的工具 ✤ 照顧參加者的需要及安全 ✤ Include representative crop varieties that have strong educational potential. ✤ Design growing space and paths for educational use. ✤ Install educational signs at prominent positions. ✤ Reserve space for storage of farming tools, stationary and equipment to support hands-on activities. ✤ Consider visitors' needs and safety.
住宅 Allotments	✤ 參閱住宅的管理指引及守則 ✤ 清楚劃分各個種植小區 ✤ 預留空間存放農具、肥料、幼苗、澆水工具等 ✤ 預留空間作公共設施,包括堆肥箱、排水及育苗設施等 ✤ Make reference to management guidelines and rules. ✤ Demarcate allotments clearly. ✤ Reserve space for stocking and installation related to supplies of tools, fertilizer, seedlings and water. ✤ Allocate space for placing communal facilities such as compost bin, drainage and nursery appropriately.
為特定人士設計的農圃 Serving specific audiences	✤ 設計時要照顧參加者的需要,例如長者需要具上蓋的休息空間;使用輪椅者需要較高的種植槽;而供兒童使用的花槽,呎吋方面也需要作出調節 ✤ Make reference to audiences' special needs in garden design. For example: sufficient sheltered, sitting areas should be designed for the elderly; raised planting beds could be designed for wheel-chair users; the dimensions of planters should be adjusted to the needs of different age groups of children.

社區農圃的基本設施
Basic Facilities for a Community Farm

設施 Facilities	作用及注意事項 Purpose and notes
工具倉 Tool store	✤ 存放農具（如鋤頭、鏟）、防蟲網、竹枝、桶等 ✤ For storage of tools (eg. hoe, spade), pest control net, bamboo poles, buckets, etc.
肥料倉 Fertiliser store	✤ 存放花生麩、骨粉、草木灰等有機肥料，可與工具倉合併，設計需能防止害蟲進入 ✤ A pest-proof, dry, sheltered area for storage of peanut cake, bone meal and wood ash. It can be combined with the tool store.
育苗區域 Nursery	✤ 育苗區可提供較佳的繁殖條件（相關資料請參閱第58頁） ✤ A nursery provides desirable conditions for effective plant propagation (refer to page 58 for details).
灌溉系統 Irrigation system	✤ 選擇合適的灌溉系統，如噴灌或滴灌（有關澆水原則請參閱第38頁） ✤ Select a compatible irrigation system, such as spray irrigation and drip-system (refer to page 38 for details).
堆肥箱 Composting	✤ 按需要選擇合適種類及大小的堆肥箱（相關資料請參閱第三章） ✤ Identify the size and type of compost bin as per the need (refer to Chapter 3 for details).
棚 Shelter	✤ 按農圃的性質及需要，設定棚的面積及位置 ✤ Determine the size and location of the shelter, based on the purpose and capacity of the farm.
廁所及洗手設備 Toilet and washing area	✤ 按農圃的性質及需要，設定有關設施的多寡 ✤ Determine the scale of these facilities based on the purpose and capacity of the farm.

農具介紹
Farming Tools

農具 Tool	用途 Application
鋤頭 Hoe	需經常使用，用於翻土及造畦。 For frequent use in tilling and ridge-making.
耙 Rake	造畦用，用作扒平田面。 For leveling and ridge-making.
鏟 Spade	開坑時使用。 For digging trenches.
園藝叉 Fork	鬆土用。 For tilling.
鏟仔 Trowel	移苗、追肥、中耕、除草時都會用到。 For transplanting seedlings, applying fertilizer, tilling and weeding.
收菜刀 Knife	收割蔬菜時使用。 For harvesting of leafy vegetables.
枝剪 Pruner	替植物修剪時使用。修枝、矮化或活動時都會用到。 For pruning plants for shading, training, stimulating growth and size control.
花壺 Watering can	市面上有不同大小的花壺，因應使用者的體力選擇合適款式。灌溉幼苗時，需使用較幼細的花灑頭。 Choose the largest can that you can manage when it is fully loaded with water. Seedlings require a fine rose.

在香港,無論是屋旁的小花園,還是露台或窗台,種植食物的地方總是不夠,因此我們應致力找出一個善用空間的種植方案。這一點,對增加農作物收成、多樣性,以至帶給個人及家庭的趣味,都是非常重要的。盆栽農圃、方塊農圃、螺旋園圃、魚菜共生及垂直綠牆等,都是具生產力而有彈性的設計,流行於本地及世界各地,值得各位城市農夫借鑒。

No matter whether you are living in a house with a garden or in an apartment with not much more than a balcony or windowsill, space for food growing food at home in Hong Kong is usually very limited. It is very important to adopt a design that makes the best use of the limited growing space for crop diversity, yield, and personal and family enjoyment. . Container-garden, square-foot garden, herb spiral, aquaponics and vertical panels are all examples of flexible and productive designs that are adopted widely by city farmers, both locally and globally, to maximize food production at home.

盆栽農圃

盆栽農圃讓家居種植變得更具彈性，容許種植者擁有更多選擇空間，例如透過搬運來爭取更多光源，或避開惡劣天氣。

窗台、露台及天台的種植設計，可與懸掛或嵌牆式的盆栽相融合，擴大種植食物的空間。不過，在盆栽種植食物，產量難免有所限制，此外，盆栽作物亦需要較頻密的灌溉。

以下是一些盆栽種植的小貼士：

材料
需防水及耐用，如玻璃、陶瓷、金屬或堅固的塑膠，但注意塑膠在陽光下損耗會較快。避免使用不耐潮或易破爛的物料。

廢物利用
車輪、木箱、膠樽及其他包裝用的廢物，都可變身成為盆栽花園的一部份。

花盆大小
農作物的體積越大，相應盆栽亦需較大。注意作物的根系形態。

Container Gardening

Container gardening is a flexible way of growing food at home. It allows food to be grown with much greater control – plants can be relocated easily for more sunlight or protection. With simple planning ahead, hanging or wall-mounted containers can be used in combination with other planting arrangements to maximize the food growing space on windowsills, balconies or rooftops. However, the size of crop is limited by the confined space and more frequent irrigation is needed.

Here are some tips for selecting containers for gardening:

Materials
Water resistance and durability are key factors for consideration. Containers could be glass, ceramic, metal or plastic. Beware that plastic may degrade quickly under the sun. Avoid containers that are made of materials which become damp or which leak easily.

Recycling
Tyres, wooden boxes, plastic bottles and other packaging 'waste' can be easily converted for use in container gardening.

Size of flowerpot
The bigger the plant, or the longer the plant takes to mature, the bigger the pot should be.

花點心思，舊物件也可以成為新花盆。
Be creative in sourcing containers for gardening.

品種
Variety

	品種 Variety	建議花盆闊度（厘米） Recommended Pot Width (cm)	建議花盆泥土深度（厘米） Recommended Depth of Soil in Pot (cm)
1	葉菜（白菜仔、莧菜、菠菜） Leaf vegetables (small white cabbage, Chinese spinach, spinach)	>30	10-15
2	葉菜（生菜、茼蒿） Leaf vegetables (lettuce, Garland chrysanthemum)	>20 （單行）(single row)	10-15
3	豆類（荷蘭豆、蜜糖豆、玉豆、綠豆） Beans and peas (sugar pea, honey pea, French bean, mung bean)	>25 （單行）(single row)	15-20
4	豆類（豆角）、青瓜、苦瓜 Beans (string bean), cucumber, bitter cucumber	>30 （單行）(single row)	>15
5	粟米 Corn (maize or sweetcorn)	>30 （單行）(single row)	>15
6	瓜類（節瓜、絲瓜等） Gourds (hairy gourd, silky gourd etc)	>45 （單行）(single row)	>20
7	茄科（番茄、茄子及辣椒） Solanaceae (tomato, eggplant and hot pepper)	>30 （單行）(single row)	>20
8	其他大型品種（白蘿蔔、青蘿蔔、紅蘿蔔、椰菜等） Other large plants (white radish, green radish, carrot, European cabbage etc)	>30 （單行）(single row)	>20
9	其他大型品種（番薯、薯仔、芋頭等） Other large plants (sweet potato, potato, taro etc)	>60 （單行）(single row)	30-40

菠菜
Spinach

茼蒿
Garland chrysanthemum

生菜
Lettuce

淺而寬的容器適合用來栽種葉菜類品種
Wide and shallow containers are suitable for growing leafy vegetables.

盆栽移植四步曲！
Four Steps for Pot planting!

1

在盆的底部鋪網，防止泥土外漏或害蟲由底部爬入。
Cover the drain holes with a fine net to block soil leakage and pest access.

2

倒入泥土，植定作物。
Fill the bottom of the container with moist soil. Transfer the plant to the container.

3

泥土的高度，應與作物的泥面保持一致。
Settle the plant with additional soil and finish it with an even surface.

4

放在有足夠陽光的地方。
Put the container in a sunny place.

小地方的實用園圃設計
Useful designs for gardening in small area

鑰匙孔式園圃

鑰匙孔式園圃設計，有助延長人與植物的接觸位置，老師或傳譯者能站在中央講解及分享，可作為教育性的園圃。花槽的闊度不宜多於一米，以便種植者兩邊均可接觸到植物。

Key-Hole Garden

A Key-Hole Garden creates a longer 'edge' for people to interact with plants, and the teacher or interpreter can stand in the middle of the planter to supervise and share. This is a practical design for an educational garden. The width of the planters should be set as less than 1m so that plants can be reached from either side.

螺旋園圃

螺旋園圃能同時提供不同的栽種環境，讓不同類別的香草能種植在同一塊小土地之上。迷迭香、百里香及其他喜歡乾旱環境的地中海氣候香草，能種於園圃的頂部。至於底部，則可種植喜歡濕潤的香草，如薄荷。至於中間位置，則可種植西蒜、羅勒、蔥等作物。為了方便管理，園圃的總闊度應控制在兩米或以下，花槽闊度亦不宜超過手臂的長度。

Herb Spiral

A Herb Spiral provides a range of growing environments to accommodate a wide variety of herbs in a small area. Rosemary, Thyme and Mediterranean herbs that require a drier growing environment can be planted at the upper area while herbs that prefer a wetter environment, such as mint, can be planted at the lower part of the spiral. The middle area can be filled with leek, basil and onion. The width of the planter should be kept within an arm's length and the overall width of the spiral should not exceed 2m.

方塊農圃

方塊農圃既便於管理，也能適應不同的環境。它讓家居種植者更有效地編排、管理一幅 4 呎 X 4 呎的小農圃，全年不匱地生產新鮮的農作物。其設立的步驟如下：

1. 選擇一幅陽光充沛的地方，開闢一個 4 呎 ×4 呎的正方形農圃。

2. 農圃四邊以磚或木板來圍繞，如地底是水泥，泥土深度應不少於 1 呎。

3. 處理好種植泥土，造好田型。

4. 以繩或線將農圃分割成 16 格，每格的面積為 1 平方呎。

5. 按照作物品種的種植距離，決定每格種多少植株（數目一般是平方數，即 1 株、4 株、9 株或 16 株）。農圃的後方設置小棚子（棚子選址不應阻擋陽光照射其他種植區域），專門種植需要搭棚的農作物。農圃的種植清單應多元化，以確保不同時期能收採到不同作物。

6. 收成一種作物後，清除其根部、添堆肥，重新播種或定植新菜苗。

7. 一般而言，八個方塊農圃（合共 128 平方呎）就可以為一個四人家庭提供足夠而穩定的蔬菜。

Square Foot Garden

Square Foot Gardening is easy to manage and very adaptable. It helps home growers to organise and manage a small garden (4 feet x 4 feet) to produce fresh produce continuously all the year round. Here are the steps for setting one up:

1. Choose a good site with full sunshine for the raised garden bed measuring 4ft x 4ft.

2. Build the four sides of the garden with bricks, recycled container pallets or wooden boards. If the chosen site has a concrete floor, the sides will have to hold soil of 1 foot in depth.

3. Fill the garden with potting soil.

4. Using string or wire to create sixteen one square foot segments.

5. Depending on the mature size of the crop, you can grow 1, 4, 9 or 16 plants per square foot. For example, if the seed packet recommends that the crops be planted 3 inches apart, you can plant 16 seeds or seedlings per square. Planting a wide variety of crops will assure a continuous harvest of different vegetables throughout the year. For crops that require a trellis for support, plant them at the back of the garden (on the side that does not block sunlight from other squares). Fix a trellis panel along that side to support plant growth.

6. Once the crops have been harvested, dig out the roots, add compost and plant new seeds or seedlings in that square.

7. Generally speaking, eight Square Foot Gardens (128 square feet) supply enough vegetables for a family of four people.

魚菜共生

建立魚菜共生的系統並不昂貴。對缺乏泥土的城市農夫來說，它是一個具彈性而有資源效益的設計。隨着水體來往魚池及種植槽之間，養份循環及水體淨化的過程能同時進行着。

Aquaponic

Aquaponics is inexpensive to set up, flexible in size and resource- efficient for city farms where soil is limited. With water circulating between the two different chambers for vegetable and fish cultivation, nutrient recycling and water purification take place simultaneously.

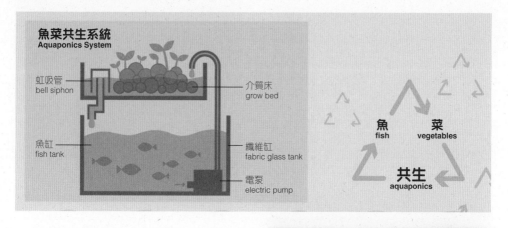

魚菜共生系統
Aquaponics System

虹吸管 bell siphon
介質床 grow bed
魚缸 fish tank
纖維缸 fabric glass tank
電泵 electric pump

魚 fish　菜 vegetables
共生 aquaponics

垂直綠牆

誰都可以造一個簡單、耐用而不昂貴的垂直綠牆。較諸分散的小種植盆形式，它提供一個較為寬闊的種植槽，容許不同類型的農作物栽種一起。綠牆的高度不應多於觸手可及的距離，這不單讓作物管理的工作變得容易，也能避免涉及沉重的硬件製造工夫。

Vertical Farming Panel

One can make a Vertical Farming Panel with easily-sourced, durable and inexpensive materials. Rather than confining individual plants in separate and small chambers, make one that provides a big common planting bed so as to allow planting a wider range of food crops. Keeping the height of panel within arm's reach not only makes crop management easier but also avoids expensive heavy-engineering work.

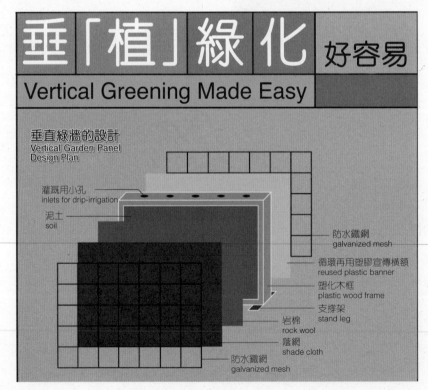

垂「植」綠化 好容易

Vertical Greening Made Easy

垂直綠牆的設計
Vertical Garden Panel Design Plan

灌溉用小孔
inlets for drip-irrigation

泥土
soil

防水鐵鋼
galvanized mesh

循環再用塑膠宣傳橫額
reused plastic banner

塑化木框
plastic wood frame

支撐架
stand leg

岩棉
rock wool

蔭網
shade cloth

防水鐵網
galvanized mesh

小水池

小水池可讓你以省水方法種植多種蔬菜，包括西洋菜、通菜、馬蹄、慈菇，甚至稻米。它並能吸引青蛙等野生小動物前來，增添生氣。

設立小水池的步驟如上：

1. 挖泥約40cm深；
2. 在池底鋪上帆布作防水層；
3. 回填泥土約35cm，設出水口以疏導多餘水份；
4. 灌濕泥土，定植作物；
5. 為水池注水。

Small pool

A small pool allows one to grow a wider diversity of crops, such as water cress, water spinach, water chestnut, arrowhead and even rice. It may also attract frogs or other wildlife.

You can construct a pool using the following steps:

1. Dig a pool of 40cm in depth.
2. Cover the bottom of the pool with tarp for water-proofing.
3. Put the soil back to the pool up to a depth of 35cm. Build in a water outlet to drain excess water.
4. Wet the soil and plant the seedlings.
5. Fill the pool with water.

在家種植！
Grow Food at Home!

像許多高人口密度的城市一樣，香港土地資源有限，食物供應的可靠度亦成疑。不過，只要善用露台、天台及其他有足夠陽光的空間，我們都可以種植食物，發展身心。

In Hong Kong, like all highly populated cities, soil is limited and food security (and safety) is always in doubt. By making wise use of one's balcony, roof or other space of sufficient sunlight, we can still grow some food to feed our body and soul.

天台種植
Rooftop farming

好的計劃是成功的一半。我們應先觀察及掌握以下一些環境因素及情況：

Having a good plan is half the way to success. Observe and take the following environmental factors into account:

風向
Wind exposure

風在高空之中的阻力較小，所以天台一般會較平地來得大風〔樓層越高，風力越明顯〕，對農作物帶來不利影響。香港出現東風的月份較多，夏季6~8月期間轉為西南季候風，入冬則轉為東北季候風。為保障作物免受強風干擾，可採用透明膠板、包書紙等材料，在向風位設置簡單、堅固的屏風設施。

A roof may be exposed to stronger winds than at ground level, as there is less friction between air and ground of wind. The prevailing wind in Hong Kong is from the east over the summer months, except for June to August when there is a shift in wind direction to the south-west. The prevailing wind is from the north-east over winter months. A strong, simple, well secured wind barrier can be set up with a transparent plastic panel or polyester film for crop protection.

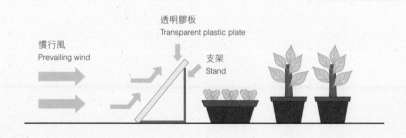

慣行風
Prevailing wind

透明膠板
Transparent plastic plate

支架
Stand

光線
Sun exposure

天台的採光量，會受到座向、鄰近建築物、地型或樹木的影響。我們要因應光資源的情況，再決定合適的種植清單。舉例説，葉菜類、茄科（如茄子、番茄和椒）及瓜類等需要較充裕陽光，豆類、香草及一些小型果樹，則可以接受略蔭涼的環境。正常情況下，農作物需要每天平均最少四小時日照，合適的光強度介乎10,000 至60,000 lux 之間。

Apart from the orientation to the sun, sun exposure of a roof garden is affected by the height of surrounding buildings, landscape and trees. Identify suitable crops to grow according to the sun exposure of different locations of the roof garden. For example, leafy vegetables, Solanaceae plants (eg. eggplant, tomato and pepper) and gourds require higher sun exposure. Beans, peas, herbs and small-sized fruit trees are more shade-tolerant. Generally crops require at least four hours of sunlight (at light intensity of 10,000-60,000 lux) per day.

牆壆
Parapet Wall

天台的牆壆，一方面可借用為農作物的「防風牆」，但另一方面它也會減少天台的日照量。我們可將農作物的盆栽位置酌量升高以加強採光；但高度應略低於牆壆，以免過於當風。採用磚等材料托高盆栽，也有助增強通風。

A parapet wall can serve as a windbreak for crop protection, and also as a barrier to block sunlight. To enjoy maximum protection from wind but suffer the least in sun exposure, planters should be raised but the height level of crops should be kept lower than the wall. Use bricks or similar things to raise the planter to increase air flow.

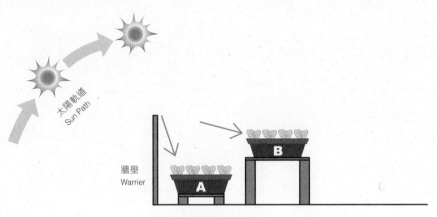

太陽軌道
Sun Path

牆壆
Warrier

A

B

B盆的日照量會較A盆多。
The daily amount of sunlight received by plot B is higher than that of plot A.

建築結構
Building Structure

規劃時，需考慮到天台的重量承載力。如涉及大型負載量，應先諮詢專業工程人士；假如未能充份掌握天台承載量情況，宜採用較為小巧的盆栽種植方案，並將重量分散於天台橫柱上方及周邊範圍，讓種植區域得到較佳的承托。

The loading limit of a roof should be considered when planning. If a large load is planned, one should ask a professional structural engineer to advise. If the loading limit of the roof is unknown, it is advisable to place the maximum load over the structural beams of the building and around the edges of the roof – in this way the load is supported by the beams and columns of the building.

作物分佈
Planting arrangement

陽光是城市農圃的最珍貴資源，我們應先確認天台的向光方向，以「梯級式」原則編排各種農作物的分佈，讓各種農作物能平均受光。

Sunlight is a valuable resource for city farming. One should design a planting plan according to the farm or garden's orientation to the sun.

圖A的攀援植物，遮擋了大部份側旁、較矮身農作物的陽光。相反，圖B的作物編排形式，能讓不同農作物較平均、有效地分享光源。

In Figure A, the climber blocks much of the sunlight from reaching the shorter crops behind. Figure B shows a different planting arrangement in which sunlight is more evenly enjoyed by plants of different height.

溫度
Temperature

水泥無論是吸熱的速度，以及貯熱的時間，均遠較泥土或植被為高。香港絕大多數的天台都是以水泥修建，夏天中午時，天台地面的溫度有機會高於攝氏50度，灼傷植物的根部。

為減低影響，夏季時種植者可利用盤架、磚塊等材料，將盆栽托高，減少與天台地面的接觸面。到了冬季，情況剛相反，「升溫」成為一個有利的種植因素，因此盆栽可直接置於地面。

Concrete is high in thermal mass and it absorbs heat faster and holds it longer than soil and vegetation. Most roofs in Hong Kong are covered with concrete and the floor temperature can reach 50°C at noon in summer. The roots of plants cannot tolerate such a high temperature.

Raising a planting pot with bricks or a shelf reduces the contact with the concrete floor and helps reduce the temperature in summer. By contrast, the warm concrete floor provides favorable growing conditions for plants in winter and therefore the bricks can be removed and pots can be placed directly on the floor.

天台上栽種食物。
Food growing on roof.

政府指引
Government Guidelines

開始天台花園前，應先參閱屋宇署就村屋製訂的適意設施公開指引，以下是部份參考原則：

承載量：總重量不超過700公斤，每平方米不超過150公斤。

園藝棚架：總覆蓋面積不超過5平方米，高度不超過2.5米。

組合櫃：高度不超過2米，體積不超過3立方米。

（資料來源：http://www.bd.gov.hk/english/documents/pamphlet/VHWUBW_b.pdf）

Before setting up a roof garden, one should take note of relevant guidelines published by the Building Department on building works. Here is a quick reference:

Loading capacity: the total gross weight should not exceed 700kg and the loading per unit area should not exceed 150kg/m³.

Garden trellis: Cover of trellis should not exceed 5m² and its height should not exceed 2.5m.

Storage cabinet: Height of cabinet should not exceed 2m and its volume should not exceed 3m³.

(Reference - http://www.bd.gov.hk/english/documents/pamphlet/VHWUBW_b.pdf)

露台種植
Balcony Garden

露台種植的編排原則，大致與天台一樣。但由於露台的採光量較少，位置較窄，設計上需作出較多的調整。

The principles for balcony gardening are similar to that of a rooftop. Adjustments are needed to cope with the limited sun exposure and space on a balcony.

光線
Sun exposure

露台的座向，對光源有決定性的影響。香港位於北半球，「朝南」的露台採光量較大，相反「朝北」就最少；「朝東」的露台採上午光，溫度較涼；「朝西」的採下午光，相對較熱。

Sun exposure of a balcony is largely determined by its orientation. A south-facing balcony in Hong Kong enjoys the best exposure to sunlight. While a north-facing balcony gets the least exposure to sunlight, especially during winter. An east-facing balcony enjoys milder and relatively cooler rays in the morning, while a west-facing balcony enjoys stronger and warmer sunlight in the afternoon

---

牆壆的修建形式
Balcony Parapet

如果露台的牆壆採用欄杆式，或以透明物料建造，由於可透光，盆栽放於較低位置亦無妨；但如果牆壆是不透光的，則應將盆栽酌量升高以加強採光。

If the parapet is made of transparent material or railing, it allows penetration of sunlight. If it is built of opaque material, there is a need to raise the planter from the shade of the parapet wall.

牆壆（透光）
Wall Barrier (Transparent)

牆壆（不透光）
Wall Barrier (Non-Transparent)

善用屋牆

Wall

露台總有一邊靠牆，如利用屋牆作垂直種植，不單能善用有限的種植空間，攀援架在牆身承托下亦會較為穩固。

If possible, make wise use of a side-wall at the balcony for vertical farming. This increases the space for farming and offers support to trellis and climbers.

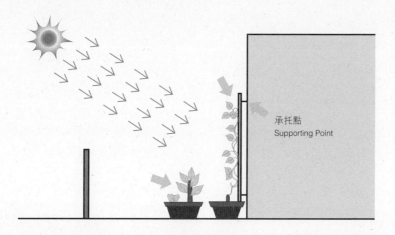

承托點
Supporting Point

以智能手機作測量工具

Mobile Apps for site assessment

今時今日，我們已可透過手機的程式，測量種植區域的不同栽種條件，包括座向、光強度、水平等實際資訊。

Today smartphone mobile apps can be used to measure orientation, light intensity and level, this is useful information for assessing growing conditions of a site.

家居種植 Q&A

問：我種植的地方很小，應該沒有需要作農圃設計吧？

答：剛好相反，狹窄的空間更需一個準確的規劃方案。舉例說，在同一個露台上，相距一米的地方，採光量可以相差一倍以上。

問：我家中空間不多，只能用花盆來種植蔬菜，那麼品種選擇上有甚麼要注意？

答：應儘量選擇較小巧的品種，例如香草、葉菜類、矮生的豆科或茄科作物(茄子、番茄和椒)等。避免選擇需要大量空間生長的作物，如果樹、部份葫蘆科(瓜類)作物等。

問：會否引來蚊蟲滋生？

答：水和泥土能滋養生命，包括昆蟲與其他動物。牠們大多數都不是害蟲，只要處理好排水問題，避免過度灌溉，蚊蟲問題應不會嚴重。

問：從園藝店見到的農具、花盆、泥土、堆肥、種子與苗，大多價值不菲。有機耕種是否一項昂貴的玩意？

答：種植無分貴賤，購買不是得到材料的唯一方法，例如不同的器皿可以循環用作花盆；咖啡渣、茶葉、廚餘等可以作為堆肥材料；種子也可以自留。這些，都是經濟而環保的替代方案。

問：我們應多久澆一次水？澆多少？怎樣澆？

答：灌溉的需要因作物而異，例如通菜喜歡濕潤，迷迭香卻需要乾旱。一般來說，淺根的作物如生菜、白菜、菠菜等，澆水需較頻密；深根作物如番茄、瓜類等，可以相隔較長的時間作灌溉。

除此之外，我們也能透過觀察天氣及表土狀況，決定是否需要澆水。

問：我怎樣才知道作物可以收成了？

答：農作物主要可分為「嫩收」和「老收」兩類。葉菜、豆(食用豆莢)及大多數的瓜類，都屬於嫩收，意即需及早收成，以避免質量因老化而下降。至於南瓜、冬瓜、老黃瓜、薑芋等，則屬於老收一類，它們需要長時間讓作物充份成長，味道、營養才會變得豐富。

Grow Food at Home Q & A

Q: Is it worth the effort to work out a farming plan for growing food in a small area?

A: A precise plan is very important to make best use of limited space in a small garden. For example, one spot's sun exposure can be double that of another spot one metre away on the same balcony.

Q: I have only a small space at home for growing crops in containers. What plants are suitable for growing under these conditions?

A: Small-size crops such as herbs, leafy vegetables, dwarf peas and Solanaceae plants such as eggplant, tomato and pepper are suitable crops for growing under these conditions. In a small garden, one should avoid fruit trees and Cucurbitaceae plants such as pumpkin that require more space to grow well.

Q: Will the farm attract mosquitos and other pests?

A: Soil and water of a farm nourish lives, including insects and other animals, but not all of them are pests. Well drained soil, avoiding over-watering and stagnant water are effective pest control measures.

Q: Farming tools, planters, soil, compost, seedlings, seeds and other supplies are expensive. Is organic farming only a hobby for rich people?

A: You do not have to buys everything you need. Reusing old containers as planting pots, collecting coffee grounds, tea and kitchen waste for compost-making, and saving and drying seeds, are earth-friendly and low cost alternatives.

Q: How often should one water, how much and using what technique?

A: Watering requirements of different crops vary. For example, water spinach is water-loving and prefers damp soil, while rosemary grows well in relatively dry soil. Generally crops with a shallow root system, such as lettuce, white cabbage and spinach, require more frequent watering. Crops with a deep root system, such as tomato and gourds, require less frequent irrigation.
Instead of fixing an irrigation routine, always observe the weather and soil condition before watering.

Q: How do I know when to harvest the produce?

A: We can broadly divide crops into 'young-harvest' and 'late-harvest'. Leafy vegetables, peas, beans and most gourds are 'young-harvest' crops – they have to be harvested in time or their quality degrades. Pumpkin, wax gourd, yellow cucumber, ginger and taro are 'late-harvest' crops – they are more nutritious and taste better if they are allowed to grow for a longer time.

屋牆可為垂直種植提供良好的支撐。
Wall offers support to trellis for vertical farming.

Chapter 02

以 永續方法 種植食物

Growing Food
in a Sustainable Way

甚麼是永續農業？
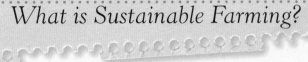
What is Sustainable Farming?

二次大戰後，慣行農業主導了廣闊的農田。化學材料的濫用、單一種植（在大範圍地段種植同一種農作物），以及工廠式的禽畜飼養系統，對環境、公眾健康、食物安全、鄉郊社區均構成莫大影響，將食物生產系統推入不可持續的方向。

Today the majority of farmland in developed countries is dominated by conventional farming – the system of chemically intensive food productive developed in the decades after World War II, featuring enormous monoculture (the practice of growing single crops intensively on a very large scale) and industralised livestock production facilities. The impacts of conventional farming on the environment, public health, food security, and rural communities make it an unsustainable way of food production in the long-term.

永續農業的綱領，就是將生態倫理與農業生產連結一起，讓我們能夠好好設計並管理農業生態系統。這系統的價值，並不限於食物生產及自然保育，也涉及文化、社會公義及經濟價值等不同層面。

The basic principle of sustainable farming is to integrate ecological rationales into agricultural production so that we work with, rather than against nature, in designing and managing agroecosystems that are both productive and natural resource conserving, and that are also culturally sensitive, socially just and economically viable.

慣行農業與永續農業比較
Comparison of Conventional Farming and Sustainable Farming

	慣行農業 **Conventional Farming**	永續農業 **Sustainable Farming**
泥土 Soil	✤ 過度依賴需大量化石燃料生產的化學肥料 ✤ 大規模開墾，改變地貌 ✤ 養份持續流失 ✤ 表土流失，土質下降 ✤ Relies on excessive application of synthetic, petroleum-based fertilisers ✤ Extensive ploughing ✤ Chronic erosion and depletion of nutrients ✤ Leads to topsoil erosion, which results in dead, compact soil	✤ 採用輪種、覆蓋植物等方法，維護及促進泥土的生產力 ✤ 循環有機資源，增加泥土有機質 ✤ 減少翻土 ✤ Promotes soil fertility and land conservation with methods including crop rotation, planting of cover crops ✤ Recycles organic resources and enriches the soil's organic matter content ✤ Minimises tillage
水 Water	✤ 化學物質和肥力的流失，對地面及其水源造成污染 ✤ Contaminates ground and surface water with toxic chemicals and fertilizer runoff.	✤ 保護珍貴水資源，採用省水灌溉方法 ✤ Conserves scarce water resources and adopts water-saving irrigation practices.
生物多樣性 Biodiversity	✤ 依賴單一種植，包括基因改造的農作物 ✤ Relies upon monoculture crop systems, including genetically engineered crops.	✤ 維護生物多樣性，對泥土保育及防止病蟲害均有所裨益 ✤ Preserves biodiversity, and it helps maintain healthy soil and keeps pests in check.
病蟲害防治 Pest and Disease Control	✤ 使用無選擇性、具毒性的化學農藥，將沒有經濟價值的動植物統統殺死 ✤ 不顧環境影響或長遠結果，只求即時取得成效 ✤ Relies on routine, indiscriminate use of toxic chemicals to kill wildlife and plants of no economic interest. ✤ Seeks an instant effect and ignores the environmental impact and long-term consequences	✤ 採用綜合式的病蟲害防治方法 ✤ 以預防方法保護農作物 ✤ 維護及加強農作物本身對抗病蟲害的能力 ✤ Adopts integrated pest management with full consideration of the overall environment. ✤ Adopts preventive measures in crop protection ✤ Emphasizes strengthening resilience of crops against pest and disease infection.

	慣行農業 **Conventional Farming**	永續農業 **Sustainable Farming**
作物 Crop	❖ 單一種植 ❖ 追求短期經濟回報 ❖ Mono-cropping ❖ Focuses on short-term, immediate economic interest	❖ 多樣化的種植策略 ❖ 重視長遠的糧食生產效益 ❖ Diversifies crop varieties ❖ Focuses on long-term resilience of crop production
市場 Market	❖ 受現行市場所主導 ❖ 追逐全球市場，作大量生產 ❖ 價格較低 ❖ 生產模式未有顧及環境及社會成本（最終要下一代承擔） ❖ Dominant in current market ❖ Usually large-scale in production that focuses on the global market. ❖ Low in crop price ❖ Environmental and social costs are externalized (Nature, everyone and future generations pay)	❖ 市場比例逐漸增加 ❖ 小規模生產，注重本土市場 ❖ 價格較高 ❖ 考慮到環境及社會成本 ❖ Growing market share ❖ Usually small-scale in production and focuses more on the local market ❖ Higher in crop price ❖ Environmental and social cost are taken into account (Charged to the customer)
化石燃料 Fossil Fuel	❖ 生產方法耗用大量能源，種植、加工、運輸過程均需依賴化石燃料 ❖ Uses energy-intensive production methods as the norm. Large amounts of fossil fuels are consumed in crop production, processing and distribution.	❖ 透過資源循環、利用本土再生能源等方法，減少對化石燃料的依賴。注重本土市場，亦有助減低食物里程 ❖ Minimizes fossil fuel use by recycling and using renewable resources that are accessible locally. Focuses more on the local market to reduce transportation mileage.

從播種到收成
From Sowing to Harvesting

① 播種

對於不少農作物，先育苗再定植至農田之中，效益會比較高。不過，仍有許多品種是適合露地播種的，以下是三種露地播種的方法：

- **撒播**：適合體積較細小的品種，例如菠菜、小白菜、莧菜等。扒平田面後，將種子平均撒在泥土表面上，再以耙輕撥田面，將種子與表土混合。幼苗長出後，需作疏苗以免過密。
- **行播**：適合通菜、潺菜、四季豆及蘿蔔等需較多空間生長的農作物。播種前，先在田畦劃上一行行淺坑，然後下種，再用泥土覆蓋種子。
- **穴播**：適合較大體積的農作物，如粟米或秋葵。先在田面按植株的距離挖一個個淺穴，每穴放二至三顆種子(種子不要連在一起)，再以泥土覆蓋。

Sowing

Whilst most vegetable seeds can be sown directly where the plants are to grow, there are some which do better when sown in a nursery and then transplanted to their growing place later, in the form of seedlings. There are three main methods for sowing hardy vegetables directly where they are to grow:

- **Broadcasting seeds:** This is the best method for growing small vegetables like spinach, small white cabbage and Chinese spinach. After scattering seeds evenly over the soil, carefully cover them with a thin layer of soil using a rake. Later, the emerging seedlings should then be thinned to the required spacing.
- **Sowing seeds in rows:** This method is suitable for crops like water spinach, basella, French bean and radish that require more room to grow. Make a shallow channel with a hoe to the correct depth (see the seed packet), sprinkle seeds sparingly in the channel and then cover with soil.
- **Sowing in holes:** This method is suitable for large crops such as okra and corn. Make holes of correct depth, put two to three seeds in each hole (keep the seeds apart) and cover the seeds with soil.

撒播
Broadcast sowing

行播
Sowing in line

穴播
Sowing in hole

播種小貼士
Sowing Tips

- 需先瞭解播種作物的株行距（每株之間所需距離），再平均播種。個別品種的資料可參閱第四章。
- 種子埋泥的深度，約相當於種子直徑的二至三倍。例如0.5cm直徑的種子，便應埋在1至1.5cm深的泥土內。
- 種子平時要存放在乾燥、清涼的地方。

- Take the planting distance of a crop (the required space between plants) into account. Refer to Chapter 4 for details about the requirements of different crops.
- Seeds should be sown evenly over the soil, at a depth which is two to three times of their diameter. For example, for a seed of 0.5cm in diameter, it should be sown at a depth of 1 to 1.5cm.
- Seeds should be stored at a dry, cool place before use.

② 澆水

適當的澆水安排，對收成會有莫大幫助。以下是幾項基本守則：

- 天氣會影響澆水的時機。一般來說，當泥面稍乾時便需澆水，而水份應滲透至根部底層。為免對作物造成傷害，應避免在最熱或最冷的時候澆水，夏季時可在早上澆水，冬季時可在中午。
- 農作物需要水份的主要是根部，不具針對性的澆水，其實也是一種浪費。另外，澆水時需考慮作物的個別需要，例如灌溉幼苗時水力不能太大，葉菜類則普遍需較頻密的灌溉；又如部份作物如番茄等，全株灌溉的話會較易誘發病害。
- 鋪設護根（覆蓋田面的材料），能有效減少水份因蒸發而流失。

Irrigation

Yield and quality of produce are greatly enhanced by proper watering. Here are some tips:

- The timing for watering changes according to weather conditions. It is better to water before drought sets in. Inspect the soil at a spade's depth. If the soil is dry, there is a need to water the crop. Avoid watering during the hottest or coldest times of the day. The best time to water in summer is early morning while the best time in winter is noon.
- Focus on the roots when watering; Beware of the different needs of different plants, for example, watering of seedlings should be made very gently to avoid damage; tomato is affected adversely by overhead watering onto the leaves, as this promote fungus and disease on leaves.
- Mulch (a layer of grass or dried leaves etc. on top of the soil) reduces surface runoff and evaporation from the soil – it saves water.

③ 疏苗

- 疏苗可讓作物有足夠的生長空間，個別品種的株行距資料可參閱第四章。
- 泥土濕度適中時，疏苗會較為容易，能避免將打算保留的幼苗也拔走。
- 疏苗後，避免立刻將植株置於烈日下，清晨時段疏苗可讓植株有較多時間恢復。

Thinning

- Thinning seedlings produces healthier plants by allowing room for proper growth and air circulation between plants. Refer to Chapter 4 for the plant distance requirements of different crops.
- Thinning while the soil is damp will help in pulling out just the excess plants, while leaving the ones you want to keep.
- Thinning at dawn gives the remaining plants time to adjust before being exposed to heat and strong sunlight.

④ 中耕

- 天雨及灌溉會令泥土逐漸變得結實。可在追肥或除草時，連帶在蔬菜之間撬鬆泥土，促進透氣。

Tilling

- Soil gets compact by regular watering and rain. During fertilizer application or weeding, slightly till the soil between the plants. This will enhance fertilization, as well as aerate the soil.

⑤ 護根

- 護根是指鋪設在田面上的可分解覆蓋物，具保水、控制雜草、長遠改善土質的作用。木糠、植物殘枝、用過的茶葉、樹葉等，都可以成為護根的材料。不過，過厚的護根也有機會引發病蟲害，一般而言，葉菜類的作物鋪設1-2cm厚的護根經已足夠。在一些情況下，也可透過種植覆蓋植物，作為田上活生生的護根。

Mulching

- Mulches are materials applied to the surface of an area of soil to conserve moisture, improve soil fertility and suppress weed growth. Many materials can be used as mulch such as sawdust, organic residues, used tea leaves, coffee grounds, newspaper and woodchip. In some situations, groundcovers are planted as a living mulch. Beware that excessive mulching can cause disease or pest problems. A 1-2cm thick mulch is generally sufficient for vegetable growing.

⑥ 修枝

對於茄科（如番茄）、葫蘆科（如冬瓜）等農作物，修枝可提升作物的生產力。個別品種的修枝要求可參閱第四章。

Pruning

Regular pruning enhances productivity of certain crops like Solanacease (eg. tomato) and Cucurbitaceae (eg. wax gourd). Refer to Chapter 4 for the pruning requirements.

⑦ 搭棚

棚子可讓種植者善用垂直空間，增加種植面積。瓜類、豆類、茄科等農作物都需利用棚子來鞏固或穩定其植株結構。

- **夏季棚**：夏季的瓜豆枝葉茂盛，因此搭建的棚子也要大一些。另外，夏季不時有颱風或暴雨，棚子的結構必需牢固。
- **冬季棚**：冬季作物如番茄、蜜糖豆、荷蘭豆等，體積上較夏季瓜豆為小，加上冬季出現暴風雨的機會較低，因此棚子的設計、用料可以較具彈性。

Trellis

A trellis adds a third dimension to a small garden and it is useful for gardeners seeking to maxmise growing space. A trellis is essential for planting crops like gourd, bean, tomato, eggplant and pepper that require support to the stem.

- **Trellis for Summer Crops:** Summer crops of gourds and beans have extensive branches and leaves. With frequent rainstorms and typhoons strong, sturdy trellises are needed for the summer-time.
- **Trellis for Winter Crops:** Winter crops like tomato, honey pea, sugar pea and other climbing legumes are smaller and rainstorms in winter are milder. Choice of trellis design and materials are then more flexible.

家居小棚子
Set up a trellis at home

都市的家居環境狹窄，難以容納大型的棚架，種植者可考慮採用「單排式」的小棚子（見圖）。

種植者可利用牆壆、屋牆等作為借力點，令棚子較為堅固。不過，小棚子的空間畢竟較小，不適宜種植一些枝葉繁茂的攀援作物。

下列是一些建議品種（以1米高「單排式」棚子為例）：

適合：青瓜、四季豆(矮生品種)、荷蘭豆/蜜糖豆、番茄

不適合：豆角、冬瓜、南瓜及其他大型葫蘆科作物

Panel trellis is more suitable than other types of trellis for home gardens in cities where space is limited. Fixing a panel trellis against a wall provides good support for planting climbers. Owing to space constraint, climbers with dense branches or leaves are not recommended for this set up.

Crops that are suitable for planting with a one-metre high panel trellis: Cucumber, French Bean (dwarf), Sugar Pea, Tomato.

Crops that are not suitable for small panel trellis: String Bean, Wax Gourd, Pumpkin and other large sized Cucurbitaceae crops.

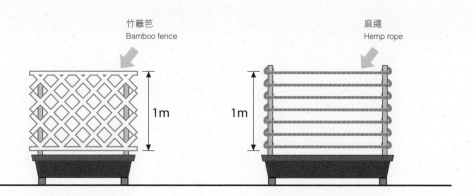

竹籬笆
Bamboo fence

麻繩
Hemp rope

1m

1m

害蟲防治
Pest Control

害蟲是一個主觀的名稱，在農田上與人類競爭農作物的生命體，便被我們稱為害蟲。

Pest is a value-laden term. Organisms which compete with human agricultural production are labelled as pests.

蝸牛
Snail

蚜蟲
Aphid

菜青蟲
Pieris rapae

黃曲條跳甲
Phyllotreta striolata

瓜蠅
Melon Fly

受雀鳥侵食的上海白菜
The Shanghai cabbage damaged by birds

慣行農民使用無選擇性、帶毒的化學農藥，將農田上沒有即時經濟利益的生物全部殺死。它只求片刻的效果，卻不會考慮對環境的長遠後果，例如：

- 生物多樣性的損失
- 污染地下水及附近水源
- 對農夫及消費者的健康構成威脅
- 對泥土的生態造成破壞，損耗地力
- 害蟲會逐漸產生抗藥性；過往有效的藥物，到後來也漸漸失去效力

Conventional farming relies on routine, indiscriminate use of toxic chemicals to kill living organisms of no immediate economic interest. It seeks an instant effect and ignores the environmental impact and long-term consequences, such as:

- biodiversity loss
- pollution of groundwater and nearby water courses
- health hazards to farmers (and possibly also to consumers)
- damage to soil ecology, which lead to soil degradation
- pest species gradually evolving pesticide resistance, which makes pest populations less susceptible to a pesticide that was previously effective at controlling the pest

永續農業對於害蟲管理的觀念，與慣行農業截然不同。與其追逐於消滅所有競爭者，不如先瞭解清楚，有關生物體與農田生態的關係，從而採取綜合性的蟲害管理方法。具體原則包括：

- 禁止使用任何合成的殺蟲水、殺草水或殺菌劑
- 對整體環境作全盤考慮，選出最合適的應對措施
- 強調採取預防的措施，與野生生物保持相對和諧的關係
- 尋求方法，加強農作物本身對病蟲害的抵抗能力
- 小心管理有機農藥的使用方法、時機及劑量，避免帶來污染，或對農民、消費者構成健康影響

The perspective of pest management in sustainable agriculture is entirely different from that of conventional farming. Rather than seeking ways to eradicate competitors, pest control measures are based on understanding of the relationships between organisms in the agroecosystems. By adopting an integrated pest management approach, sustainable farming:

- prohibits use of synthetic chemical pesticides, herbicides or fungicides;
- takes full consideration of the overall environment in selecting the best pest control methods to address an identified pest problem
- emphasizes protective measures that harmonize the relationship with wildlife
- seeks ways to strengthen the resilience of crops against pest and disease infection
- manages the timing and dosage of applications of organic pesticides to avoid potential pollution and health hazards to farmers and consumers

預防方法

- 不時不植，強壯的農作物自然有較高的抵抗能力
- 收成後適時休耕、曬田，利用陽光殺死病菌
- 採用輪種方法以保持地力，並中斷病蟲害在泥土的繁殖周期
- 採用間種方法，並增加農作物的多樣性，以建立平衡的生態系統，抑制病蟲害爆發
- 種植能夠驅趕害蟲的植物，如萬壽菊及薄荷
- 移除病源，及時清理腐爛的生果及其他作物
- 潮濕環境容易帶來病害，可透過調節灌溉來避免泥土過濕
- 選擇適合當地氣候、抵抗力較強的品種

Preventive Methods

- Grow according to seasons yields stronger crops, which are more resistant to pests and diseases.
- Leave farmland fallow for a while after the main crop is harvested for it to restore fertility and to let sunlight control soil-borne diseases and pests.
- Adopt crop rotation to break the reproduction cycles of pests and diseases and prevent them from establishing in the soil over time. Crop rotations also help maintain soil fertility.
- Adopt intercropping and increase the diversity of crops to establish an ecological balance for pest control.
- Plant pest-repelling plants such as marigold or mint.
- Remove ripe fruit and rotten produce quickly.
- Regulate watering and monitor soil moisture. Plants are more susceptible to pest and disease infection under an over-humid environment.
- Source seeds and seedlings of crop varieties that are adaptable to the local climate and more resistant to local pest and diseases.

物理防治

- 菜青蟲等害蟲，可以人手清除
- 農田上懸掛眼型汽球、反光物料或風中會發聲響的驅趕工具，以嚇走雀鳥
- 不能使用會令雀鳥受傷的鳥網。鳥網的絲很幼，很容易纏着雀鳥或蝙蝠。如不能及時將網移走，會導致這些動物重傷或死亡

Physical Control

- Pest like caterpillars of Small Cabbage White can be picked and removed manually.
- Use visual methods to scare birds, such as scare-eye balloons, reflective materials that flash in the sunlight or tools that make noise in wind,.
- Never use mist-net for crop protection as it is relatively invisible and the fine netting tends to trap wild animals like birds and bats. Unless the trapped animals are rescued immediately with great care, being caught will result serious injury or even death.

不會傷害雀鳥的保護網。
Protective net which is bird-friendly.

生物防治

- 生物防治是引入有關蟲害的天敵及病害，以達到針對性的防治效果。
- 自然界中，生物防治其實四處可見。青蛙及蜻蜓，都是生物防治的好例子。

Biological Control

- In pest management, biological control is a method of managing pest populations by introducing or inviting parasites, predators or diseases into the pest's environment, to the detriment of a particular pest population.
- Biological control occurs in nature all around us everyday. Frogs and dragonflies are effective biological control to restrain insect pests.

天然農藥

蒜頭水、辣椒水及魚藤液，都是以天然材料製造的有機農藥。

Natural Pesticides

Garlic juice, pepper juice and derris are made of natural materials for use as pesticides in organic farming.

魚藤
Derris

自製瓜蠅誘捕器
Melon Fly Trap DIY

瓜蠅是農田中一種主要害蟲，對許多瓜類構成傷害，包括苦瓜、冬瓜、青瓜、水瓜、南瓜、西瓜、翠玉瓜等。自製的瓜蠅誘捕器，可有效地作出防治。

Melon flies have more than 80 hosts. They are major pests of bitter melon, Chinese wax gourd, cucumbers, luffa, pumpkins, watermelon, and zucchini. Melon fly is one of the most important pests with which vegetable growers have to contend. A homemade melon fly trap is easy to make and effective against flies.

材料

所需材料非常簡單，包括一個容量約500毫升的透明膠樽、一塊膠枱墊（約20cm X 30cm）、約300毫升的石榴汁（或以爛瓜果替代）及一卷烏蠅紙。

製作步驟

1. 在水樽上半部份割開多個十字形開口，並向內掘而成孔，每個直徑約4厘米。
2. 在枱墊上開一個孔，套入水樽的頂部。
3. 在水樽底部貼上烏蠅紙。
4. 將石榴汁或爛果放入樽內。
5. 懸掛誘捕器在田中，石榴汁可吸引瓜蠅前來，牠們接觸膠樽時會被烏蠅紙黏着，或浸死水樽內。

Materials

The only supplies you need to make a homemade melon fly trap are a 500ml transparent plastic bottle, a 20cm x 30cm plastic sheet, 300ml of juice (rotten fruit or gourd can be used as substitutes) and a roll of fly paper.

Procedure

1. Make four holes on the upper part of the bottle by cutting multiple crosses and pressing them in; each hole with a diameter of 4cm.
2. Poke a hole at the centre of the plastic sheet and fit it on the top of the plastic bottle.
3. Stick the fly paper at the bottom of the plastic bottle.
4. Put juice or rotten produce into the bottle.
5. The homemade melon fly trap is ready for use. Hang it in the garden, flies will be attracted to it and go into the bottle. Once they are trapped inside the bottle, they are either stuck on the fly paper or drown in the liquid.

註：為保持功效及避免出現衛生問題，每隔三至四星期便要更換果汁。
Replace the juice and the fly paper once every three to four weeks in order to maintain its effectiveness and hygiene.

木醋液

木塊在低氧高溫處理的炭化過程中，釋放出來的煙經冷卻後而成的液體，主要成份包括有機酸及酚類物質。木醋液本身有防菌、防蟲、除臭、活化微生物等作用，適用於農業及園藝。在一些國家如日本等，木醋液是一種普遍的農業材料。

由於木醋液不帶損害人類健康的物質，也不會導致環境污染，是一種生態友善的驅病材料。

功效

病蟲害防治：　　木醋液本身可以抑制病菌害蟲、保護葉面，對於潮濕天氣、通風不足而出現的病蟲害，包括綫蟲、立枯病、蚧殼蟲、潛葉蛾、鏽病等，具有一定功效；對幼苗效果尤其明顯。

提升植株的活力：將木醋液施用在農田、堆肥以至植株上，有助提高有益微生物的數量，加強植株根部及葉面的活力，提升農產品的質量。

用法

- 使用木醋液，應先以清水稀釋100-150倍，然後一併施用在植株及其土壤之上。（註：有關比例是按嘉道理農場暨植物園生產之木醋液為參考依據，如使用其他來源之木醋液，比例或需作出調節。）

- 避免在下雨及烈日的情況下施用，以免減低成效。

- 木醋液的酸度很高(pH 3 - 4)，未被稀釋前，應避免直接接觸皮膚。此外，不要將木醋液與鹼性的農藥混合使用。

Wood Vinegar

Wood vinegar is the dark liquid collected from the condensation of the smoke that is released when wood is heated in the absence of oxygen during the charcoal making process. Its application in horticulture and agriculture include bacteria and insect control, deodorization and activation of soil microbes. Wood vinegar comprises of mainly organic acids and phenols and it does not contain any polluting substance that is harmful to humans. It is regarded as an earth-friendly alternative for crop protection and is widely applied in agriculture in countries like Japan.

Functions

Pest and disease control – Wood vinegar suppresses the spread of disease and pests on leaves. It works well in controlling disease and pests associated with humid weather and weak ventilation, such as nematodes, damping-off, scale insects and leaf-miners. Its performance on seedlings is particularly significant.

Activate plant growth – Applying wood vinegar on farming soil, compost and plants activates growth of beneficial microbes, plant roots and leaves, that enhance crop productivity and quality

Application

- Dilute wood vinegar with water in 1:100 to 1:150 ratios before application on soil and plants. (This is the mixing ratio for the wood vinegar produced by Kadoorie Farm and Botanic Garden. For use of wood vinegar produced by other sources, there is a need to adjust the mixing ration as per the instruction specified.)
- To assure the best performance, avoid applying wood vinegar when it is raining or the intensity of sunlight is very high.
- Wood vinegar is highly acidic (pH3 - 4). Avoid contacting undiluted wood vinegar and do not mix it with any alkaline pesticide.

參考資料：台灣國立宜蘭大學有機產業發展中心《木醋液之製造及使用》
Reference – Production and usage of wood vinegar, Organic Center of National Ilan University

稻草人

Scarecrow

稻草人常見於有機農場。它不會傷害雀鳥，目標只是將牠們趕到其他地方，不再成為農場的麻煩製造者。今時今日，稻草人已成為一個生態友善的象徵。

Because they do not harm the birds, they are widely used in organic gardens where they discourage birds from feeding on crops, while the birds may feel comfortable to be in other parts of the farm where their presence is not a problem. Today, the scarecrow is being seen as an icon of eco-friendliness.

農田中有一個稻草人，總是一件好事。它不單協助保護農作物，也為四周帶來了歡樂。其實，稻草人不一定要用「稻草」來做，外表也不一定是「人」型。要設計一個有效的稻草人，首先要代入侵害者的角色，如雀鳥，想像牠們究竟會害怕甚麼，以及甚麼東西會令牠們離開農田。

It is great to have scarecrows in a garden, not only because they help protect crops, but they are also fun to make and have around. A scarecrow is not necessarily made of straw nor is it necessary to have it in a human-shape. To design an effective scarecrow, you have to firstly identify the target pest. Then put yourself in the role of the pest and imagine what elements can deter your from eating the crops.

以下是一些參考的構思：

1. 顏色：採用鮮艷色彩或反光的物料
2. 聲音：製作一些會在風中發出聲響的工具
3. 搖擺：懸掛一些會隨風擺動的物件，如色帶
4. 特別圖案：製作一些類似動物眼睛的圖案，以阻嚇雀鳥

Here are some ideas for reference:

1. Visual scares: bright colours and/or reflective materials that flash in the sun
2. Sound scares: something like a mobile with metal parts that move freely and make sound in the wind
3. Mobile scare: something made of light strips of material that sway in the breeze
4. Special pattern: patterns that looks like the eyes of animals are effective in driving birds away

註： 野生動物如雀鳥等，面對外在環境時具備一定適應性，一些陌生東西牠們剛見到時會懷疑，但時間一久，
便可能不再害怕。因此，稻草人的外觀及位置需不時作出轉變。
Wild animals like birds are wary but adaptable. They steer clear of anything that looks suspicious or out of place,
however, they get used to it if it stays for a while. In order for your scarecrow to work, it needs to be changed and moved
around from time to time.

有機農藥DIY
Organic Pesticide DIY

蒜頭及辣椒水

這些驅蟲材料對害蟲有一定阻嚇作用，且易於製造。

- 辣椒水：將30克的辣椒搗爛，混入1公升的清水中
- 蒜頭水：將100克的蒜頭搗爛，混入1公升的清水中

製作完成後，噴灑在農作物的葉面上。

Garlic and Pepper Sprays

These juice sprays are easy to make and effective in deterring pests.
- To make pepper juice spray, mix 30g of smashed pepper with one litre of clean water.
- To make garlic juice spray, mix 100g of smashed garlic with one litre of clean water.

For application, spray the mixture on the surface of crop leaves.

啤酒陷阱

在潮濕或下雨天，經常有許多蝸牛跑來吃菜的葉片。你可裝設啤酒陷阱來解決這問題。

1. 將容器半埋入泥
2. 注入啤酒
3. 啤酒的酵母及碳水化合物能吸引蝸牛，並跌入陷阱中
4. 定時檢查容器，並清走死蝸牛
5. 啤酒不足時，及時補充

一個陷阱已足夠一個小農圃使用。如面積較大，可同時裝設數個陷阱

Beer Trap

Snails feed on the leaves and roots of plants, and they are very active on humid, rainy days. You can address the problem by setting a beer trap in your garden.

1. Fit a watertight container (eg. The bottom part of a plastic bottle) in soil with its mouth at ground level.
2. Fill the container with beer.
3. Snails are attracted to both the yeast and the carbohydrates in the beer. They will fall into the cup and drown.
4. Check the beer trap daily and remove the dead snails.
5. Refill the bear when it evaporates to a low level.

A single trap is sufficient for a small garden. For a larger garden, you can place several traps around the perimeter of the garden.

驅蟲植物

例子一：馬利筋 / 連生桂子花

馬利筋能吸引蚜蟲的天敵瓢蟲過來，從而控制蟲害。此外，它亦能吸引金斑蝶等蝴蝶，令園圃更具生氣。

Pest Repelling Plants

Example 1: Blood flower (*Asclepias curassavica*)

Blood flower attracts aphid as well as its predator, ladybug for pest control. It also serves as the food plant for Plain Tiger (*Danaus chrysippus*) which in turn make the garden lively.

例子二：萬壽菊

萬壽菊的氣味有助驅走蚜蟲及蚊子，根部則能夠驅趕綫蟲。它是茄科作物的好拍檔，同時具有很高的觀賞價值。

Example 2: Marigold (*Tagetes patula*)

The scent from marigolds repels aphids and mosquitos. The roots of marigolds repels nematodes (roundworms). It is a good companion plant to tomato and pepper. It is also highly ornamental.

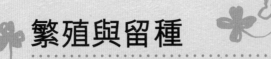

繁殖與留種
Seed Propagation and Seed Saving

農作物的繁殖,主要依靠「有性」(透過種子)及「無性」(透過插枝、壓條、分株、嫁接等)兩個方法。

Plant propagation is the process of creating new plants either by sexual (from seeds) or asexual propagation methods (eg. Cuttings, layering, division and grafting).

種子繁殖

大多數蔬菜以種子繁殖下一代,但其留種形式各自不同,好像豆角、玉豆、蜜糖豆、番茄、生菜等屬於自花授粉;白菜、蘿蔔、莧菜、青瓜及芹菜等,則屬於異花授粉。

Seed Propagation

Most vegetables are propagated by seeds, but the forms and conditions for seed sowing and saving vary. Crops like string bean, French bean, sugar pea, tomato, eggplant and lettuce are mostly self-pollinated (flowers fertilized with pollen of the same plant) while crops like white cabbage, radish, Chinese spinach, cucumber and celery are cross-pollinated (flowers of one plant fertilized with pollen of another plant).

異花授粉的莧菜
Variety of cross-pollination:Chinese Spinach

留種原則

留種是一門老技藝,農民選擇自己喜歡的品種,一年一年保留下去。以下是幾項留種的基本原則:

- **選種**:選取作為留種的植株,應該是強壯而無病的。種植者可以抗病能力、生長速度、健康情況、外貌等作為挑選的標準,剔出不合留種規格的植株,在開花前將其清除。

- **隔離**:自花授粉的作物對此問題不大,但對異花授粉的農作物來說,留種時要避免近親(如節瓜和冬瓜)栽種在附近,以免出現雜交。

- **數目**:對於自花授粉的作物(如番茄或生菜),即使只預留小量植株,仍可以順利留種。但對於異花授粉的農作物如白菜或莧菜,則必需保留一定植株數目(由數株至數百株不等),以確保作物能夠順利完成授粉。

Principles of Seed Saving

Seed saving is as old as agriculture. Throughout history farmers have considered seed from their favorite plants to be treasures well worth saving from year to year. Here are a few tips for seed saving:

- **Selection:** Identify your favorite plants, which are strong and healthy, to be the breeding plants for seed saving. Resistance to pests and diseases, growth rate, health condition, quality and yield are the major criteria for selection.
- **Isolation:** Seed saving of inbreeders (pollinates itself) is relatively easy. To save seeds of outbreeders (cross-pollinates with other plants), there is a need to isolate the seed plants from other varieties of the same genus (such as hairy gourd and wax gourd) to prevent pollens from different varieties being able to reach each other and cross-pollinating the varieties.
- **Quantity:** For inbreeders (eg. tomatoes and lettuces), only a small number of seed plants are needed for seed saving. For outbreeders (eg. white cabbage, Chinese spinach), a few dozen of seed plants are required to assure successful pollination.

無性繁殖

採用無性繁殖的農作物(包括番薯、薑、芋、分蔥等)具備以下特點：

- 下一代的基因與原體完全一樣，因此有利複製優良植株；
- 生長速度較種子繁殖快；
- 只有少數植株亦能繁殖；
- 植株的繁殖沒有經過自然環境的篩選，因此不能因應環境變化而作出調節。當遇上環境轉變或特別的病害，較容易出現集體死亡。

Asexual Propagation

Crops propagated by asexual methods have the following characteristics:

- The offspring are genetically identical to the parent plants, and therefore, advantageous traits can be preserved.
- It propagates and yields faster
- Only one parent plant is required
- The plants gradually lose their vigour as there is no genetic variation. They are more prone to diseases that are specific to the variety, which may result in the destruction of an entire crop.

維護品種多樣性

粉嶺的鶴藪白菜、打鼓嶺的雷公鑿苦瓜、川龍的西洋菜，都是昔年非常出名的新界菜品種。過往，農民會按照當地的環境情況，將這些品種一代一代保存下去。

可以，現代農業促使全球化的新品種不斷出現，而它們大多是雜交而來的，這導致傳統品種逐漸流失，而農民對種子商的依賴則越益加深。這些雜交品種的產量的確可能高一些，但它們都是不適合留種繁殖的。因此，選擇本地品種並自行留種，有助維護本地的傳統品種，以及農作物的多樣性。

Preservation of Agricultural Diversity

Hok Tau white cabbage from Fanling, bitter gourd from Ta Kwu Ling and watercress from Chuen Lung were once well-known local varieties of agricultural crops. These are varieties developed by local farmers over generations and they adapt well to the local environment.

Following the Green Revolution, (most notably a big push to increase agricultural output from the 1960's onwards, using new technologies such as widespread use of chemical fertilizers and pesticides, and hybrid seeds) the push to 'modernise' agriculture promoted rapid global extension of 'modern' varieties, mainly hybrids, which have been developed by commercial seed companies by crossing specific parent plants. This haslead to a rapid loss of traditional varieties and an increase in farmers' reliance on costly commercial supplies of seeds and synthetic fertilisers and pesticides, which usually come as a package. Today, seeds in the market are often hybrids – they may yield higher than traitional varieties, but their seeds are often sterile, or do not reproduce true to the parent plant. By choosing to grow local varieties and by practicing seed saving, famers and gardeners can help to preserve traditional varieties and the overall agricultural diversity, in which the resilience of agro-ecosystem is rooted.

種子的貯藏
Seed Storage

種子的活力會隨時間而減退，如貯藏情況乏善，今年使用的種子，到第二年發芽率已經會明顯下降。乾燥、低溫及黑暗，是貯放種子的基本條件。下列程式可作為貯藏種子的參考指引：

$$相對濕度（\%）+ 溫度（℃）\leq 50$$

家居種植者，當然不會有專業的貯藏儀器。但可參考以下方法保存種子：

- 貯藏前，先確保種子乾爽
- 按品種包好，列明品種及年份
- 將種子與乾燥劑一同放入雙重的密封膠袋中
- 貯放於雪櫃下層（約2-5℃）

The viability of seed decreases over time, but poor storage conditions greatly shorten the viability. Dry, cool and darkness are the basic conditions for seed storage. The following formula illustrates the relationship between temperature and humidity requirements for seed saving:

$$Relative\ Humidity\ (\%) + Temperature\ (℃) \leq 50$$

For home growers to keep seed viable for a longer period in the absence of professional equipment, there is a need to:

- Make sure the seed is completely dry before storage.
- Pack seed with a paper envelop and label it with the variety and year.
- Place the envelopes in an air tight container or a zip-lock bag together with a pack of desiccant
- Store the container in a refrigerator under a temperature of 2-5 ℃.

綠薄荷的插杆方法
Procedure for Cutting Spearmint

選擇老身的枝葉。
Select strong mature stems.

每段保留約10厘米,將底部的葉除去。
Each stem should be kept about 10 cm long. Clear the leaves of lower part stem.

插入適合插枝的土壤中。
Put a few stems into a pot.

定時澆水,頭數天避免讓陽光直接照射。
Water regularly. Avoid direct sunlight in the first few days.

培苗技術
Nursery

為甚麼要培苗？
Why is a nursery necessary?

育苗的基本作用，是讓農作物的幼苗階段可在一個較佳環境下成長。一個理想的育苗區，應能發揮下列的效果：

- **節省種子**：較佳的育苗環境，有助提高種子的存活率及植株長勢。
- **節省農地**：育苗所需的空間只佔露地田的1-5%。如果將農作物的前期生長階段集中在育苗區，農地在同一時間便能栽種更多作物。
- **減少管理工作**：集中育苗，能提高管理效率，並節省時間及用水。
- **提早收成期**：育苗區的播種時間，一般可以較露地播種提早半至一個月，換言之能夠提早收成。
- **統一幼苗規格**：種子發芽，難免會出現參差。育苗可讓種植者在定植農作物之前先作編排，及早淘汰弱株，提升農地的有效使用率。

A nursery provides desirable conditions for effective plant propagation. Ideally, a nursery should contribute to the following:

- **Save seeds:** By providing a desirable environment for plant propagation, it enhances the seed generation rate and produces stronger seedlings.
- **Save farming space:** A nursery occupies only 1 – 5% of the farmland area. By grouping propagation and early growth of different crops within a nursery, more farmland is released for production.
- **Enhance management:** Grouping propagation of different crops at one place enhances management efficiency in terms of time and irrigation.
- **Extending production period:** Compared to sowing seed directly outdoors, a nursery allows sowing to take place by two weeks or even one month earlier, and therefore extends the production period.
- **Quality assurance:** There is always variation on germinating performance between seeds. Propagation at a nursery allows farmers to select strong seedlings for transplantation, which enhances effective use of the farmland.

培苗的主要考慮因素
Core factors for propagation

- 育苗環境（光線、溫度、通風）
- 苗土質量（疏水、通氣、保水保肥能力）
- 育苗容器（大小尺寸、帶穴或不帶穴）
- 作物的特性（植株體積、根系再生能力）

- Environmental factors (light, temperature, ventilation)
- Soil quality (drainage, ventilation, capacity to hold water and fertility)
- Seedling container (size, with cell or without cell)
- Characteristics of individual crop (size of crop, root regeneration ability)

育苗環境及時機
Environment and timing for propagation

不同季節的育苗工作，重點會有所不同。在華南地區，蔬菜育苗的高峰期集中於春播（2、3月）及秋播（8、9月）。

The focus of propagation work differs between seasons. In Southern China, the peak times for propagation are spring (February and March) and autumn (August and September).

	春播（2月-3月） Spring sowing(February to March)	秋播（8月-9月） Autumn sowing(August to September)
重點 Focus	儘量保溫及透光 Maximize penetration of sunlight and heat insulation.	減低暴雨、強烈光線的影響，促進通風 Block damage from rain and strong sunlight; enhance ventilation

	春播（2月-3月） **Spring sowing(February to March)**	秋播（8月-9月） **Autumn sowing(August to September)**
育苗棚設計 Nursery design	✤ 密封或半密封 ✤ 儘量採用透光的材料 ✤ Fully or partially covered ✤ Use transparent shelter to maximize sunlight penetration	✤ 上蓋擋雨 ✤ 外圍不宜密封，以免烈日下溫度過高 ✤ 上頂可能要掛遮蔭網，避免烈日直曬 ✤ Shelter from rain ✤ Do not fully cover the periphery of the shed for temperature regulating ✤ Cover the top with shade-net to filter strong sunlight
苗架高度 Height of nursery	不宜貼地，最好能升高一些，免受積水、害蟲、雜草等影響 Put the nursery bed above ground level to protect seedlings from water-logging, pests and weeds.	
育苗時機 Timing	立春、雨水（2月初至3月初）期間仍不時有寒流，宜於苗棚內進行育苗。過了驚蟄（3月初），露地育苗亦已合適。 During 'Start of Spring' and 'Spring Shower' (February) of the Solar Terms, cold monsoon weather is still frequent and propagation is more successful if carried out at nursery. After 'Insects Waken' (March), sowing directly outdoors becomes feasible.	秋分（9月下旬）前仍常有颱風暴雨，宜在棚底育苗。品種方面，時機大致如下： ✤ 8月中起：白菜、菜心、芥菜、番茄、茄子 ✤ 9月初起：生菜、油麥、茼蒿、椰菜、西蘭花、椰菜花、芥蘭、芥蘭頭、君達菜、芹、蒜、秋播香草 ✤ 9月底：荷蘭豆、蜜糖豆、豆苗 Typhoon and rainstorm are frequent before "Autumn Equinox" (late September), propagation within a nursery is preferred from mid-August until then. The best time for sowing different varieties of crop: ✤ From mid-August – white cabbage, flowering cabbage, mustard leaf, tomato and eggplant ✤ From early September – lettuce, Indian lettuce, garland chrysanthemum, European cabbage, broccoli, cauliflower, Chinese kale, kohlrabi, Swiss chard, celery, garlic and autumn-sowing herbs ✤ Late September – sugar pea, honey pea and pea seedlings

苗土管理
Potting Soil

苗土要兼顧保水肥、通氣、作物適應性等因素。農友可以向外購買部份材料，混合農場的田泥而成為育苗土。現時市面常見的苗土材料有以下幾類：

When choosing potting soil, factors like water and fertility-holding capacity, aeration and its compatibility with the chosen crops should be taken into account. One can purchase essential materials and mix them with farm soil to prepare potting soil. Materials that can be purchased for potting mix:

珍珠岩

受熱膨脹之岩石，密度低、輕、透氣、疏水。雖然效用良好，但因成本較貴，生產過程涉及的能源亦較高，建議節省、適量使用。

Perlite

This volcanic rock product is lightweight and low in density, which enhances aeration and drainage of soil. Application of perlite should be minimal as it is expensive and consumes high energy to produce.

河沙

同樣具透氣、疏水作用，效果不及珍珠岩顯著，但成本遠為便宜。需要留意的是，育苗必需採用河沙，不能使用鹽份太高的海沙。

River sand

It enhances aeration and drainage of soil. Although its performance is less effective than perlite, its price is much lower. Beware that only river sand, not sea sand can be used due to the salinity of sea sand.

堆肥

本身帶肥力，亦能改善苗土質量。

Compost

It acts a fertilizer as well as a soil conditioner.

泥炭土

苔蘚分解而來的有機物質，能有效改善土質、保水保肥。但由於原材料屬自然界有限資源，成本亦高，故不宜濫用。

Peat moss

Peat moss forms when mosses and other living materials decompose in peat bogs. It is a soil conditioner which holds moisture and nutrients in soil. Since it is an expensive, non-renewable natural resource, application should be minimized.

除此之外，亦有人會使用**椰衣、蛭石**等作為苗土材料。種植者可視乎其結構形態，適量調配。

Other materials such as **coconut fibre and a group of minerals named vermiculite** are also used as potting mix. The mixing ratio should be adjusted according to the nature and character of different materials.

混合比例
Mixing ratio

由於苗土原材料的品質時有參差，因此調配的比例需具有彈性。一般來說，混合好的苗土應具備以下特徵：

- 顏色偏暗黑
- 手握起來，略具彈性；不會有大量沙石在手指隙間瀉下
- 苗杯盛載苗土後加水，水份不致積存在苗土上久久不散

The mixing ratio should be adjusted flexibly according to the type and nature of raw materials. Good potting soil should possess the following features:

- dark in color
- resilient in texture – not much soil particles slips through fingers when it is gripped
- no sign of water logging after watering

混合好的苗泥
Potting soil mix

建議比例
Suggested mixing ratio

如果育苗土以田泥為基礎，由於田泥土質較佳，因此可以使用較少外來物料，參考比例是（以容積計算）：

田泥50%、泥炭土20%、堆肥10%、珍珠岩10%、河沙10%

不過，如果沒有田泥可用，便惟有購買外來的黃沙花。黃花沙相對貧瘠，需混合較多其他材料作為輔助，參考比例是（以容積計算）：

黃花沙35%、泥炭土35%、堆肥10%、珍珠岩10%、河沙10%

Farm soil can be used as the base of potting soil mix. Farm soil is often good in quality and therefore less external input is needed. The suggested mixing ratio by volume is:

50% farm soil, 20% peat moss, 10% compost, 10% perlite, 10% river sand

Yellow sand mud can be used to substitute farm soil if the latter is not available. As yellow sand mud is often poor in quality, more external materials are needed for potting soil preparation. The suggested mixing ratio by volume is:

35% yellow sand mud, 35% peat moss, 10% compost, 10% perlite, 10% river sand

育苗器皿
Propagation containers

育苗容器主要分「帶穴苗盤」、「不帶穴苗盤」、「苗杯」三類。

Propagation containers can be broadly classified as **tray with cells, tray without cells and seedling pot.**

帶穴苗盤
Tray with cells

格數越多的苗盤，苗的體積越小。現時較常用的有50格、72格等。帶穴苗盤所育的苗，移植時能連泥帶走，對幼苗根部損傷較少，復原亦較快。苗穴呎吋方面，大苗穴有較大空間，不易盤根，緩衝期長，效果當然較小苗穴好，但材料及管理成本會相對較高。

50-cell and 72-cell trays are most common. The larger number of cells that a tray contains, the smaller the size of each cell and thus the seedlings that it can hold. Trays with cells allow seedlings to be transplanted with the potting soil attached, which causes less root damage during transplantation. Larger cells provide more room for root growth and longer buffering time for transplantation. Larger cell trays are better but cost more and require more input.

不帶穴苗盤
Trays without cells

苗盤不設獨立苗孔，幼苗移植時不能連泥帶走，對幼苗損害較大，復原期較長。但好處是能在有限空間培育較多數量的幼苗。

For trays without cells, seedlings are transplanted without soil, therefore risking greater damage to the seedlings. The advantage of a tray without cells is that more seedlings can be propagated in the tray at the same time.

苗杯
Seedling pots

直徑由2吋至4吋不等。以獨立容器育苗，效果當然較為理想，但管理成本亦較高，適合用於特別品種或對外活動。

A seedling pot is usually 2 to 4 inches wide. Propagating seedlings with seedling pots yields better results but it is higher in cost. It is suitable for propagating special varieties or serving special uses in educational activities.

不同類型的苗盤、苗杯。
Different types of propagation tray and pot.

帶穴盤所育幼苗，定植時能連泥一併移植，避免根部受損。
For trays with cells, seedlings are transplanted with the soil attached; this reduces root damage.

品種特性

Characteristics of different crops

不同型類的蔬菜，育苗的方法亦會有所不同。我們應充份掌握作物的生長特性，然後制定合適的育苗策略。一般來説，育苗前應考慮以下因素：

1. 植株大小

植株越大，育苗效益越明顯。所以，體積較大的作物(如番茄、瓜類、椰菜)適宜用以帶穴苗盤育苗；體積較小的作物(如白菜、生菜)就可採用不帶穴苗盤。

2. 生長期

生長期越長，育苗的效益亦越大，例如香草。

3. 根系再生能力

菊科(包括生菜、油麥、茼蒿)及十字花科(包括白菜、芥菜、菜心、芥蘭)的根系再生能力甚強，即使以不帶泥苗定植亦問題不大。相反，藜科(包括菠菜、紅菜頭、君達菜)、傘形花科(包括芹、荽、大小茴)等農作物，根部受傷後不易恢復，因此忌用不帶穴盤。

4. 對於收採根部的農作物，如蘿蔔、甘筍等，幼苗定植的步驟可能會導致主根分叉，因此最好還是露地播種。

5. 部份品種即使採用露地種植，成活率亦相當高，所以未必需要育苗，如豆、通菜、潺菜、薑芋等。

No propagation method suits all crops. Choose the appropriate method according to the need and characteristic of the respective crop. Take the following factors into account when identifying the propagation method to use:

1.Size of crop

The larger the plant, the benefit of seedling propagation in a nursery is more obvious. A tray with cells is suitable for propagating large-sized crops such as tomato, gourds and European cabbage. For small-sized crops such as white cabbage and lettuce, a tray without cells can be used for propagation.

2.Growing period

The longer the growing period of a crop (eg. herbs have a long growing period), the bigger the benefit of seedling propagation in a nursery.

3.Root regeneration

Asteraceae (eg. lettuce, Indian lettuce and garland chrysanthemum) and Brassicaceae (eg. white cabbage, mustard leaf, flowering cabbage and Chinese kale) have strong root regeneration ability and therefore the impact of transplanting without soil is not significant. However, the root regeneration ability of Chenopodiaceace (eg. spinach, beet root and Swiss chard) and Apiaceae (eg. celery, coriander, fennel and dill) is poor and therefore they should be propagated in a tray with cells and transplanted with the soil.

4. Transplantation of root-vegetable seedlings (eg. Chinese radish and carrot) may damage the main root and lead to poor harvest. Seeds of this type of crop should be sown directly outdoors.

5. Crops like pea, water spinach, basella, ginger and taro have a high germination rate even if seeds are sown directly outdoor. There is no need to propagate these in a nursery.

低成本苗圃設計！
Set up a nursery with low cost

精耕細作形式
Adapting an existing shed

以現成苗架改建，棚的上、後方蓋以溫室膠紙，以遮擋風雨；左、右、前方則採用24目防蟲網圍起，前方可揭起，護苗之餘亦具透氣效果。夏季及初秋時宜將前簾打開，避免棚內溫度過高。

An existing shed can be modified to a seedling nursery. Cover the top and back side of the nursery with plastic sheets to shield wind and rain. Cover the two sides and the front with net (hole size as 24 holes to 1 inch). Make the front net flexible so that it can be opened for temperature regulation. The side-net should be open in late summer and early autumn for ventilation.

膠箱小苗室
Plastic nursery box

將塑料魚缸或透明麵包箱等，改成為小型溫室。記着必需在上方設通風小洞，避免育苗區溫度太高。這方法適合用於市區小農圃，又或天台、露台種植等。

An old plastic fish tank or transparent bread case can be easily modified to be a mini-greenhouse. There is a need to drill holes at the upper part of the box for temperature regulation. This design suits a small community garden, rooftop garden or balcony garden.

農耕日曆
Farming Calendar

太陽是主導地球氣候的最重要因素，因此，瞭解陽曆（根據太陽黃道編排的曆法）的規律，能協助我們掌握四時變化。

中國的二十四節氣，便是一個按陽曆方法編排的農耕曆法。每一個節氣，都代表太陽在黃道某一個位置，農夫能按着節氣來編排一年之間的農耕工作。

現今二十四節氣的名稱，始自於西漢時期（公元前206年至公元後24年）。黃河流域是中華民族農業的發源地，二十四節氣描述的景物，也是以黃河流域為依歸。由於南北之間氣候存有差異，因此華南地區套用二十四節氣時，詮釋上會與北方有所差異。

The sun is the most influential single element that determines the weather on earth. From the Earth's perspective, the sun moves along a path known as the ecliptic over the year.

In China, the farming calendar organizes a year in the format of 24 solar terms. The 24 solar terms divide the ecliptic into 24 equal segments. Each term signifies the relative position of the sun in the ecliptic at a certain times of the year. The calendar helps farmers to stay synchronized with the seasons in organizing their work.

The origin of the farming calendar with 24 solar terms can be traced back to West Han (206BC-24AD) at the cradle of Chinese agriculture, the Yellow River in Northern China. The 24 solar terms reflect the climate over central China in ancient times. There are minor discrepancies between Northern and Southern China regarding the descriptions of weather, so the Hong Kong farmer must make adjustments for these.

二十四節氣的編排
The 24 Solar Terms

廿四節將地球環繞太陽的軌道，劃分成二十四份，每個季節有六個節氣，每個節氣之間相距14-16日。每個季節的第一個節氣均以「立」字為首，即「立春」、「立夏」、「立秋」和「立冬」；而每個季節的中間（第四個節氣），則是「分」或「至」，即是「春分」、「夏至」、「秋分」和「冬至」。

「春分」和「秋分」是晝夜均等的日子，「夏至」日照最長，「冬至」則最短。其他節氣的名稱，很多是描述當時的天氣情況（例如雨水、清明、小暑、大暑、處暑、白露、寒露、霜降、小雪、大雪、小寒、大寒），又或與農務活動有關（如穀雨、小滿、芒種）。

The 24 solar terms divide the ecliptic, the path which the sun moves through a year, into 24 equal segments. There are six terms per season and the period between terms is 14-16 days. The name of the first term in each season begins with 'Start' (Start of Spring, Start of Summer, Start of Autumn and Start of Winter) while the fourth term of each season ends with Equinox or Solstice (Spring Equinox, Summer Solstice, Autumn Equinox and Winter Solstice).

At Spring Equinox and Autumn Equinox, the periods of daylight and night are equal in length. The period of daylight is the longest on Summer Solstice and the shortest on Winter Solstice. This was the first set of solar terms, determined in ancient China. Other solar terms were named according to the weather (eg. Spring Shower, Clear and Bright, Moderate Heat, Great Heat, End of Heat, White Dew, Cold Dew, Forest Descends, Light Snow, Heavy Snow, Moderate Cold and Severe Cold) and farming activities (Grain Rain, Grain Forms and Grain on Ear) that are prevalent at the respective times of the season.

為甚麼要採用陽曆？

Why do we adopt the solar terms in the farming plan?

陽曆是按太陽軌跡(其實是地球軌道)來計算，與日照、降雨、溫度及風向等主要種植條件，都有直接關係。相對之下，陰曆是按月球軌道來推算，其相關之條件主要在於引力及晚間光度，重要性相對低一些。

不過部份農業理論系統如生物動力農法(Biodynamic)等，在編訂農耕工作時，亦會兼採陰曆作為參考。

The Chinese 24-solar terms track the orbit of the Earth around the Sun to assess the various factors for farming, such as sunlight, rainfall, temperature and wind direction. On the other hand, the Chinese lunar calendar is counted according to the orbit of the Moon around Earth. It mainly reflects conditions like gravity and luminosity at night, which is less relevant to farming activities.

However, for some agriculture theories like Biodynamic farming, the lunar calendar is used to organize agricultural work.

具體的節氣日子如下：
The details of the Chinese 24 Solar Terms:

	節氣 Solar Term	新曆日期 Date in Western Calendar	意義 Implication
1	立春 Start of Spring	2月4或5日 4th or 5th of February	冬盡春來，大地回春。為提高農耕效率，一些農夫已在溫室等地方進行育苗，以便充份掌握即將來臨的種植良機。 Winter ends and spring arrives. To increase productivity, some farmers have already begun nursery sowing to prepare for the coming good growing season.
2	雨水 Spring Shower	2月19或20日 19th or 20th of February	「雨水」是指雪轉為水。香港雖然沒雪，但這時的濕度及溫度亦會逐漸上升。 Snow melts and water comes. There is no snow in Hong Kong but the humidity and temperature gradually increase.
3	驚蟄 Insects Waken	3月5或6日 5th or 6th of March	春雷乍響，蟄伏的昆蟲也被驚醒。這時春耕工作應已全面展開。 Spring storms break and waken dormant insects. All spring farming should have begun.
4	春分 Spring Equinox	3月20或21日 20th or 21st of March	第一個晝夜均等的日子。天氣和暖，應充份把握春耕的時機。 The first instance of equal length of night and day. As the weather is warm, farmers should grasp the opportunity to farm.
5	清明 Clear and Bright	4月4或5日 4th or 5th of April	清明多雨，滋潤植物，因此生機蓬勃。 A rainy term. Plants get irrigated and become lively and vibrant.
6	穀雨 Grain Rains	4月20或21日 20th or 21st of April	雨生百穀，種植作物的好時機。 Rain nourishes grain. It is golden opportunity to grow crops.
7	立夏 Start of Summer	5月5或6日 5th or 6th of May	夏天開始。除了種植，也要為防澇等夏災工作做好準備。 The beginning of summer. Flood prevention should be done alongside the gardening.
8	小滿 Grain Forms	5月21或22日 21st or 22nd of May	小滿是指穀物種子逐漸飽滿。 Grain Fills means that the seeds of grain become mature.
9	芒種 Grain on Ear	6月5或6日 5th or 6th of June	穀物成熟。天氣在此時已越來越熱。 The crops are fully grown. The weather gets steadily hotter.
10	夏至 Summer Solstice	6月21或22日 21st or 22nd of June	白晝最長的一天，象徵一年最熱的日子即將來臨。 The longest day, signifying the arrival of the hottest days of the year.
11	小暑 Moderate Heat	7月7或8日 7th or 8th of July	非常炎熱的日子。由於溫度高、害蟲多，風暴又頻密，對香港農夫來説，這是艱苦的時候。 Torrid weather. For Hong Kong farmers, this is a difficult time because of the high temperature, frequent rainstorms and pests outbreaks.
12	大暑 Great Heat	7月23或24日 23rd or 24th of July	非常炎熱，種田苦守的時候。 Very hot, waiting for better weather.

	節氣 Solar Term	新曆日期 Date in Western Calendar	意義 Implication
13	立秋 Start of Autumn	8月7或8日 7th or 8th of August	香港在這段時間仍然很熱，不過，如細心觀察，便會發現入夜時會逐漸轉涼。這時也是農民積極準備秋播的時刻。 Still very hot in Hong Kong but you may notice that the weather becomes cooler after dusk. Time for farmers to prepare for autumn sowing.
14	處暑 End of Heat	8月23或24日 23rd or 24th of August	天氣雖仍熱，但暑氣已漸衰，這時秋播工作應已展開。 Although the weather is still hot, the heat of summer is fading away. The autumn sowing should have started.
15	白露 White Dew	9月7或8日 7th or 8th of September	天氣漸涼，水凝而成露，已是種植的理想時候。 The weather becomes cooler and dew appears. This is a good time to plant.
16	秋分 Autumn Equinox	9月23或24日 23rd or 24th of September	第二個晝夜均等的日子，許多秋冬蔬菜這時都適合種植了。 The second instance of equal length of night and day. A good time to plant autumn and winter vegetables.
17	寒露 Cold Dew	10月8或9日 8th or 9th of October	天氣清涼，且暴雨減少，是蔬菜生長的良好時候。 Cool weather with fewer storms. A good environment for vegetables to grow.
18	霜降 Frost Descends	10月23或24日 23rd or 24th of October	黃河流域這時已出現降霜，香港也到了深秋時分。 The Yellow River Region is white with frost while it is already late autumn in Hong Kong.
19	立冬 Start of Winter	11月7或8日 7th or 8th of November	冬季開始。但這時環繞香港的，仍是濃濃秋意。 Winter begins but Hong Kong is still enveloped in late autumn weather.
20	小雪 Light Snow	11月22或23日 22nd or 23rd of November	香港不會真的下雪，但氣溫會持續下降，也要留心寒流，以免農作物受破壞。 It does not snow in Hong Kong but the temperature drops. Beware of cold snaps which are harmful to crops.
21	大雪 Heavy Snow	12月7或8日 7th or 8th of December	香港的冬季仍然適合耕種，但也要做好防寒準備。 Hong Kong's winter is still a suitable time to plant while measures to prevent cold damage to crops should be taken.
22	冬至 Winter Solstice	12月21或22日 21st or 22nd of December	白晝最短的日子，一年最冷的日子開始來臨。 The shortest day of the year, signifying the arrival of the coldest days in a year.
23	小寒 Moderate Cold	1月5或6日 5th or 6th of January	這時種植的，都是一些耐寒的蔬菜。 Plants which grow in this period are hardy.
24	大寒 Severe Cold	1月20或21日 20th or 21st of January	寒冬之最，象徵春季將臨。防寒之餘，也是時候規劃未來一年的農耕工作。 Late winter, meaning spring is coming. While protecting the crops from cold weather, it is time to draw a plan for the agricultural work next year.

Chapter

03

泥土篇
Soil
Management

好食物，由好泥土開始
Good food starts with good soil

泥土的形成與成份
Soil Formation and Composition

泥土的主要成份是礦物質及有機質。我們必需緊記，並非所有泥土都適合用於種植食物。現時作為農業用途的表土，其實是經過數以千年時間的風化及生物作用，才逐點逐點累積而成的。可惜的是，現時慣行農業並不重視土壤護理，任由泥土外露而遭受水土流失，許多經過漫長歲月才累積起來的薄薄淺土，在短短十數年間便損耗掉。前車可鑒，所有栽種者都應將保護、改善土質作為首要任務。

此外，泥土也是一個蚯蚓、細菌等匯集而具有生命力的地方。我們應將泥土及其內在生物視為一個整體，不要讓燒田、施藥、破壞性機械、單一種植等不合適的方法，將豐沃的土壤變成光禿禿、沒有生命力的泥巴。

The basic components of soil are minerals and organic matter but not all soil is suitable for food growing. It takes thousands of years of weathering and biological activities for the topsoil, a vital agricultural resource, to establish. However, soil preservation is often ignored by today's conventional farming, and mismanagement of agricultural land is one of the main causes of soil degradation worldwide. In view of this, all growers should rank soil preservation and enrichment as their top priority.

Soil is also the home of earthworms, micro-organisms and many other living creatures. Don't let fire, agrochemicals, improper farming methods such as use of heavy farm machinery and monocropping undermine the vitality in the soil.

改善土質的管理方法	損耗地力的管理方法
✓ 採用循環有機廢物、種植綠肥等方法，保持、提升泥土的有機質含量	✗ 只重視泥土的養份（過度依賴施肥），忽視土壤的結構（缺乏補充有機質）
✓ 利用覆蓋植物、護根等方法，減少泥土暴露空氣的面積及機會，降低風雨侵蝕的影響	✗ 任由農田的泥土廣泛暴露，造成嚴重水土流失
✓ 採用迂迴去水道設計，讓雨水沖走的泥土可以在田間沉澱，以便重新使用	✗ 過度翻土，破壞泥土的結構，加速泥土有機質的損耗
✓ 善用護根、免掘園圃等方法，減少翻土的次數，降低對泥土結構造成的破壞	✗ 單一種植、濫施農藥，又不進行休耕，年復年的過度吸收泥土內部份養份，讓土地沒有休養生息機會
✓ 採用休耕、輪種等方法，保持地力	

Soil enriching practices	Soil depleting practices
✓ Continuously enriching organic matter in soil by recycling organic waste and growing green manure in soil ✓ Protecting soil from erosion by wind or rain by mulching and planting groundcover ✓ Establishing circuitous water channels to allow more time and distance to retain sediment in the farming field ✓ Maintaining the integrity of the soil structure by mulching, no-tillage and other methods that minimize turning of soil ✓ Adopting rotation practices, including letting the land rest and lie fallow for a period to restore its fertility	✗ Relying on agrochemicals to replenish the nutrient content in soil. Neglecting the preservation of soil structure and replenishment of organic matter in soil ✗ Exposing soil extensively to water and soil erosion ✗ Excessive tillage destroys soil structure and intensifies erosion of top soil ✗ Monocropping leads to imbalance in the soil nutrients by letting the crops draw the same kind of nutrients from soil year after year, on the same land. Farmers may also become increasingly dependent on pesticides, as particular types of pests proliferate

香港的土質
Soil in Hong Kong

泥土的有機質含量，受溫度、降雨、植被等因素影響，香港位處亞熱帶，溫暖、潮濕的環境，促使微生物活動、有機質分解、植物生長等情況變得較為活躍，泥土中能累積的有機質比例會低一些。如果種植地點本身是農地，土質一般較有保障。但如果是新開墾地(尤其是坡地)，土質便可能較遜。

The organic situation in any one soil is to a large degree an expression of temperature, precipitation and vegetation. In general, the rate of microbial activity, decomposition of organic matter and plant growth is accelerated with increasing temperature. As Hong Kong is located in a sub-tropical region, the warm and humid climate favors rapid decomposition of organic materials. Therefore, organic matter content in our soil is relatively low. Soil quality is generally better in areas where there was formerly farmland. Soil quality is inferior in areas that have been newly opened up for farming, especially if it is located on a slope.

如何鑑別土質？
What is soil texture?

砂土、壤土及黏土
Sand, Silt or Clay

泥土的結構，以其顆粒的大小來分辨。砂質土的顆粒較大，排水快，不利保存水份及養份，黏土則反之。不同的砂、壤、黏土質配搭，構成了不同類型的泥土。

認識泥土對編排種植來說是非常重要的，如要瞭解本身泥土屬於甚麼類型，可透過一個簡單實驗來確認：在泥土中加一些水，握一把在手中，然後將泥土慢慢搓成長條狀。如果是黏土，會較容易握成條狀；如果是砂土，未成為長條狀時應早已崩裂。

Soil texture refers to the size of particles that make up the soil. Sand, having larger particles, drains easily but its ability to retain moisture and nutrients is poor. Silt, being moderate in particle size, has a smooth or floury texture. Clay, having smaller particles, feels sticky and has high ability to retain moisture and nutrients. The various combinations of sand, silt and clay result in different soil textures. A good understanding of soil texture is essential for designing a planting plan. Here is a simple experiment for assessing soil texture: add water to soil, grip it and try to rub it into a strip. Soil with high clay content can be rubbed into strip easily, while soil with high sand content tends to break before a strip is formed.

有機質
Organic matter

有機質含量較高的泥土，一般呈暗黑色，手握泥土時會感到有點彈性；如果泥土呈淡黃、淡紅，手握時沒有彈性或立即崩散，則顯示其有機質含量較低。

Soil with high content of organic matter is usually dark in colour and spongy in texture. If the soil is yellowish or light red in colour, and falling into pieces when it is gripped, it suggests the soil contains a fairly low level of organic matter.

家居種植，泥土從何而來？
Where do we source soil for use in a home garden?

這是許多家居種植者首要面對的問題。有些人會乾脆在園藝店鋪購買科學泥或育苗土，但這些材料的價錢、生態足印均很高，不值得鼓勵；此外育苗土本身保水保肥能力太強，也不利作物根部的長遠發展。

從園藝公司購買的黃花砂
Yellow sand mud from a horticultural store

一般的折衷做法，是從花園、園藝店等，購買俗稱「黃花砂」的泥土。這些泥土相對便宜，但土質較差，所以，我們需加入一些改善土壤的物料，例如堆肥、泥炭土、珍珠岩、河沙等等〔土壤改良物料的特性，可參閱P.61〕。混合的比例隨物料質量、種植品種而調節。正常情況下，我們建議以大約60% 的黃花砂作為基礎，再混合其他改良物料。如果是購買培苗土，也可採用60%黃花砂、40%培苗土的比例。〔以容積計算〕

This is the first question to be asked by most home growers. It is costly both financially and environmentally to buy synthetic soil or potting mix from a horticultural store. Added, the ready-to-use potting mix's water and nutrient retention ability is too high and can hinder root growth.

An alternate is to purchase the cheaper 'yellow sand mud' from horticultural stores as the base for soil preparation. There is a need to improve its quality by mixing in soil enriching materials such as compost, peat moss, perlite and river sand. (refer to P.61 for details). The mixing ratio should be adjusted according to the quality of mixing materials and the planting plan, for example, Chinese radish and rosemary prefer sandy soil while water spinach and basella prefer clay soil. Generally the soil-mix contains 60% of yellow sand mud as the base by volume. One may prepare the soil-mix by mixing **60% yellow sand mud with 40% of ready-to-use potting soil** purchased from the market.

家居種植小貼士：實用材料家中尋！
Tips for sourcing useful materials for a home garden

不少人會花許多金錢，購買各種不同形式的土壤改良劑。但你有沒有想過，這些東西其實可能是唾手可得，而又不費分毫的呢？

許多可分解、不帶污染性的植料，都可按比例與泥土混合，作為土壤改良劑或肥料。例如：

- 咖啡渣
- 茶葉（花茶如杭菊、香草等亦可）
- 盆栽修剪下來的枝葉

如果植料體積較大，宜先輾成小段，才再與泥土混合。過硬或附着污染物的材料則不宜採用。

One can buy everything needed for one's garden or reuse available materials at home for soil enrichment. The latter costs nothing.

Plant and other non-polluting biodegradable materials can be mixed with soil as soil conditioner or fertilizer. Here are some examples:

- Coffee grounds
- Tea leaves, including scented tea like chrysanthemum or herb
- Pruning-waste of stems and leaves

Chop plant waste into small pieces before mixing with soil. Avoid hard or contaminated materials.

免掘農圃
No Dig Garden

免掘種植
No-dig Garden

翻土雖然具備鬆土、帶氧的正面作用，但過度翻土，也會引致水土流失、泥土結構受損的不良影響。

而免掘種植，是仿做自然界的運作，以覆蓋方法減少水肥、土壤的流失，並能抑制雜草生長及促進有機資源循環。其步驟如下：

1. 以園藝叉將田泥撬鬆；

2. 上面鋪上容易分解的綠色植料，如菜筴、嫩草等；

3. 蓋上一層堆肥，充份澆水；

4. 以濕透的報紙完全覆蓋田面，再鋪上護根；

5. 大約一星期後，便可撥開護根，戳破報紙，將菜苗定植於田中。

Tillage can loosen the soil and bring benelitial oxygen into it. However, frequent tillage can lead to negative impacts including soil erosion and destruction of soil structure.

The "No-dig Garden" system imitates the operating system of nature. It relies on a thick cover to prevent soil erosion, stops weeds from growing and enhances organic recycling in soil. The steps of preparing a No-dig Garden are as follows,

1. Loosen the soil with a garden fork.

2. Make a layer of green plant material which is easily decomposed; for example vegetable leftover and fresh grass.

3. Lay compost and water thoroughly.

4. Use water-drenched newspapers to cover the entire surface of the ridge. Lay mulch on top.

5. Transplanting can be done after about 1 week. Slightly brush the mulch aside and poke a hole through the newspaper. Place the vegetable seedling inside the hole.

有機肥料
Organic Fertilisers

有機肥料的基本概念

- 必須保持泥土的養份均衡。
- 施肥的時間和份量，要與環境和植物配合，切忌過度施用！
- 設法保持泥土養份不流失。
- 施肥時要兼顧泥土的結構。

種植者需掌握營養元素的特質，才能做好肥力管理的工作。不過，植物涉取的養份元素共有十六種之多，要一一牢記並不容易（事實上也無必要）。有一個相對簡單的記誦方法，就是「3+1」。

3 是指「主要元素」，即氮、磷、鉀；

1 指「微量元素」，包括鐵、硼、鈣、銅、鎂等等。

Basic concepts of organic fertilisers

- Maintaining a balanced set of nutrients for the soil is a must.
- Fertilizer application timing and quantity should match the environmental conditions and the maturity of plants. Do not over-fertilise!
- Make every effort to retain the soil nutrients.
- The structure of the soil has to be duly considered in fertilization.

Having a good understanding of the soil nutrient profile guides the way to good soil fertility management. Among the sixteen essential nutrient elements for plant growth, one can focus on the **"3 + 1" formula**:

"3" refers to essential elements, which are nitrogen, phosphorus and potassium.

"1" refers to trace elements that include iron, boron, calcium, copper, magnesium etc.

養份元素 Element	有機肥例子 Organic source	作用 Purpose	肥力特徵及施用方法 Characteristic and application
氮 Nitrogen	✤ 花生麩粉、魚肥、咖啡渣 ✤ Peanut cake, fish scrap and coffee grounds	✤ 促進枝葉生長 ✤ Promote leaf growth	✤ 氮肥移動性強，易流失，例如施用一遍花生麩粉後，肥力大抵只能維持10-20天。 ✤ 因此，施用氮肥需配合時機，最好能把握農作物前期、枝葉繁密生長的階段。 ✤ 較適宜以追肥方式施用。 ✤ 過多會妨礙開花。 ✤ Nitrogen-based fertilizer is highly mobile in soil. For example, fertility retention time after application of peanut cake is about 10-20 days. ✤ The ideal time for fertilizer application is during the early stage of plant growth. ✤ Apply as top-dressing. ✤ Excessive nitrogen hinders flowering.
磷 Phosphorus	✤ 牛骨粉、牛蹄甲 ✤ Bone meal (cattle), Hoof meal (cattle)	✤ 有助開花結果 ✤ Promote root growth and flowering	✤ 磷肥穩定性強，不易流失。例如施用一次骨粉後，肥力可維持數個月。 ✤ 較適宜以基肥方式施用。 ✤ Phosphate fertilizer is stable and stays in soil for months after application. ✤ Apply as a base fertilizer before sowing or transplantation of seedlings.
鉀 Potassium	✤ 草木灰、礦物鉀肥、紫草水 ✤ Plant ash, potash fertilizer and comfrey water	✤ 促進植物內碳水化合物的合成與傳送 ✤ Help plant synthesize carbohydrates and regulate metabolic activities	✤ 如果使用的是草木灰，其穩定性較低、易流失，需小心貯藏及配合時機使用。 ✤ 如果使用的是礦物肥，其穩定性較高，可作基肥使用。 ✤ Plant ash has low stability and easily drains out. Avoid application on rainy days and store it in dry area. ✤ Potash fertilizer is relatively stable and can be used as a base fertilizer before sowing or transplantation of seedlings.

養份元素 Element	有機肥例子 Organic source	作用 Purpose	肥力特徵及施用方法 Characteristic and application
微量元素 Trace elements	堆肥 Compost	❖ 所需份量不多，但長期缺乏會影響植物吸收，並易生病害 ❖ A deficiency in any trace element in soil can limit plant growth even when all other essential elements are present in adequate amounts.	❖ 由於所需量低，不宜輕率獨立施用。否則可能破壞泥土本身養份平衡。 ❖ 有機質的分解過程中會釋出多種微量元素。只要我們定期施用堆肥，便無需太擔心微量元素不足。 ❖ Only required in small quantity. ❖ Excessive content of individual trace element may upset nutrient balance in soil. ❖ Degradation of organic matter releases various types of trace elements. Sufficient supply of trace elements can be provided by regular application of compost.

常見的有機肥料
Common organic fertilizers

花生麩粉
Peanut Cake

骨粉
Bone meal

草木灰
Ash

綠肥
Green Manure

家居肥料大搜查！
Fertilizers that are readily available at home!

1. **豆渣**：氮肥含量高，但需小心避免採用基因改造的黃豆。施用時，應與表土混合，避免過份集中招來害蟲。
2. **咖啡渣**：施用方法與豆渣大致相同。
3. **蛋殼**：收集時，宜先以清水沖走蛋漬，然後涼乾、壓碎，這樣的話可有效保存數個月時間。收集後，以樽密封保存數天後才使用，效果更佳。
4. **洗米水**：在第一次洗米時，刻意放少一些水，這樣收集時可提高養份濃度。
5. **植物枝葉**：修剪農作物、園藝植物時剩下的枝葉，只要未木質化、沒病害，也可切成短截混入泥土中。

1. **Soya residue:** This has high nitrogen content. To avoid pest nuisance, mix it well with top soil and spread it out well during application. Avoid application of residue from genetically-modified soya.
2. **Coffee grounds:** Apply in a similar way as soya residue.
3. **Egg shell:** Wash it after collection. Dry and crush it before application. It can be stored for several months.
4. **Recycle water after rice washing:** Use less water for the first round of rice washing and collect the used water which is concentrated with nutrient. It can be stored in a jar for a few of days before application to enhance the performance.
5. **Pruning-waste of stems and leaves:** Select those that are disease-free. Non-woody materials decompose faster.

蛋殼
Egg shells

咖啡渣
Coffee grounds

施肥小貼士！
Tips for fertilizer application

- 平衡施用，缺少了某一門養份，不能以其他的養份元素取代。
- 定期使用堆肥、護根，提高泥土有機質的比例，這樣有助保存養份不易流失。
- 大雨不施肥。
- 多年生作物在休眠期間不施肥。例如木瓜、菠蘿等熱帶作物不宜在冬季施肥。
- 作物已呈老化時，施肥的邊際效用會下降；有時種植新株會更為划算。
- 肥料與表土混合，讓根部更易吸收。

- Beware of nutrient balance and pay attention to deficiency of certain elements.
- Regular application of compost and mulching enriches organic matter content in soil gradually, which improves the nutrient-holding capacity in the long run.
- Never fertilize when there is heavy rain.
- Do not fertilize perennial crops (eg. pineapple and papaya) when they are dormant.
- Replanting new crops is more cost-effective than fertilizing aged crops after a major harvest.
- Mixing fertilizer into top soil enhances the plants' nutrient uptake.

肥料建議使用量（以 $10m^2$ 面積菜田計）
Under a general situation, a $10m^2$ garden requires

- 花生麩 1KG
- 骨粉 1KG
- 堆肥 10KG

- peanut cake / 1KG
- bone meal / 1KG
- compost / 10KG

你也可以自製堆肥液！
Liquid Fertiliser DIY

液態肥料能讓植物更快吸收，但也較容易流失於表土層。為提高效益及預防造成污染，液肥可作為短線的養份補充材料。

Plants quickly take up fertilisers in fluid form but liquid fertilser also leaches quickly into the subsoil. To make effective use of resources and avoid pollution, liquid fertilizer should be applied only as a short-term measure to address nutrient deficiency.

花生麩水
Liquid Peanut Cake

- 將1公斤花生麩粉，放入15-20公升的清水中。
- 花生麩粉分解期間，會傳出非常難聞的氣味，因此所用的水桶必需以蓋密封。
- 分解過程最少需兩、三個月，麩粉會變成糊狀，所浸的麩水亦會變得混濁。為免造成肥傷，不要使用未分解成熟的麩水。
- 完成後，將麩水稀釋8至10倍，施用於植株根部。

- Put 1kg of peanut cake in a bucket and add 15-20 litres of water.
- It stinks when the peanut cake decomposes. Tightly cover the bucket with a lid.
- It takes two to three months for the peanut cake to decompose fully to a mush of muddy liquid fertiliser. Never apply immature fertilizer to plant to avoid damage.
- Dilute the liquid fertilizer in a ratio of 1:8 to 1:10 before application. Apply it to the roots of plants.

紫草水
Comfrey Water

紫草含豐富的蛋白質及鉀元素，施用紫草水對瓜、豆、果類等農作物均有顯著效果。

- 將1公斤切碎的紫草葉放入桶中，注入12公升的清水。
- 封蓋。分解過程需4至6星期，這時的紫草水會呈棕黑色。
- 施用於植株的根部。

Comfrey is rich in protein and potassium, and therefore, comfrey water is particularly good for growing crops like gourd, beans and fruit. To make comfrey water:

- Put 1kg of chopped comfrey in a bucket. Add 12 litres of water.
- Cover the bucket with a lid. It takes four to six weeks for the comfrey to decompose fully to brownish black liquid fertilizer.
- Apply comfrey water to the roots of plants.

紫草
Comfrey

紫草水
Comfrey Water

有關堆肥
Something about Compost

為甚麼做堆肥？
Why Make Compost?

自然界內，生物體在死亡後會被微生物、真菌、無脊椎動物所分解，成為較簡單的形態，回歸至生態系統的養份循環之中。「有機廢物」這概念在自然界其實並不存在。

然而人為活動的介入，令自然界的養份循環受到干擾。根據環境保護署的資料，香港平均每天棄置於堆填區的家居廢物，高達6,418噸（2014年資料），當中約有三成是有機物。

將這些「有機廢物」改變成促助當地園圃及農場的有機資源，這是一個轉廢為能的好方法。

當中的過程，更可以提高參加者對環境責任的認識。

In nature, bodies of living organism begin to decompose shortly after death. The organic substances in invertebrates, fungi, plants and micro-organisms are broken down into much simpler forms of matter, and recycled in nutrient cycles. There is nothing called 'organic waste' in nature.

However, human interventions have created gaps and blocked the natural nutrient cycles. According to the Environmental Protection Department, Hong Kong dumped 6,418 tonnes of domestic waste in landfill every day in 2014, and one-third of this was categorized as 'organic waste'.

Imagine if the 'organic waste' was recovered to be organic resources for nourishing local gardens and farms. The process converts a waste problem into a resource for cultivating abundance.

Compost-making at community farms not only generates resources from waste, the process itself offers educational opportunities to increase environmental awareness and arouse curiosity.

社區堆肥可提高參加者減少廢物的意識。
Community composting can increase the participants' awareness of waste reduction.

家庭種植推薦使用！

堆肥基本理論
Composting Basics

製作堆肥，就是提供良好的條件，讓生物能有效地分解各種有機物質。

現時，市面上的堆肥主要以禽畜糞便、木糠等材料來製作。不過，我們可以儘量收集一些不受污染、容易分解而又較易在生活上取得的材料。以下是一些例子：

To make good-quality compost effectively, one has to provide favourable conditions for organisms to decompose organic substances.

Although compost products in the market are usually made of livestock manure and sawdust, there is a wide choice of materials that can be used for making compost. Look out for organic matters that are uncontaminated, decompose quickly, and are easy to source and collect. Here are some examples:

原材料 Raw Material	注意事項 Remarks
廚餘 Kitchen waste	✤ 未經煮食的材料，如菜筴、蛋殼、果皮、瓜皮等 ✤ 為保障衛生，避免使用骨頭、肉類、帶油或調味料等材料 ✤ Uncooked waste such as vegetable trimmings, egg shells, peelings of fruit and gourds are easier to manage. ✤ Avoid bone, meat, oily and seasoned substances, for hygienic concern.
茶葉/茶包 Used tea leaves or tea bags 咖啡粉渣 Used coffee grounds	✤ 許多餐廳及咖啡室會樂意提供這些材料 ✤ Many restaurants and cafes are willing to supply these materials.
農作物殘株、雜草 Crop residue	✤ 避免採用已開花或帶種子的雜草，以免其再度生長 ✤ But avoid flowering or seed-bearing weeds – these may germinate later.
園藝植物枝條 Horticultural waste	✤ 較粗的枝條需要長時間來分解，不宜大量採用；又或先將其切至小段 ✤ It takes a long time for wood to decompose. Cut wood into small pieces before composting and limit the amount in use at one time.

* 為免引起生物安全的問題，不應使用人或寵物的糞便。

For biosecurity concern, do not use manure of pets or humans in compost making.

廚餘
Kitchen waste

枯葉
Dried leaf

影響製造堆肥的因素
Factors Affecting Composting

除了尋找原材料，我們也要設法提供適合堆肥運作的環境條件。當中影響堆肥製作的六個最主要的基本因素：

氧氣、碳氮比、材料表面積、
溫度、水份及酸鹼度

Apart from selection of raw materials, favourable conditions should be provided to assure production of high quality compost the six primary factors which affect composting are:

**Oxygen, carbon-to-nitrogen (C:N) ratio,
surface area of materials, temperature,
water and acidity/alkalinity level (pH value)**

氧氣
Oxygen

好氧的生物以呼吸維持生命，合適的通風環境可令分解過程更迅速，並且不帶臭味。促進通風的方法包括：

- 堆肥箱的設計需預留空隙，確保空氣流通
- 定時翻動堆肥
- 不要擠壓堆肥物料
- 較細密的堆肥材料，需與較粗的材料混合
- 將管子插入堆肥，增強透氣

Aerobic organisms need to breathe to survive. Aeration is necessary in aerobic composting to assure rapid, odor-free decomposition of organic substances. To enhance aeration,

- The compost bin should be designed with some space inside to enhance aeration
- Aerate a compost pile by turning it
- Do not compact the composting materials
- Thoroughly mix composting materials of various sizes
- Insert aeration pipes into the compost pile

碳氮比
Carbon-Nitrogen (C:N) ratio

碳和氮是微生物的生命支柱，合適的碳氮比例可以提升堆肥材料的分解效率。理想的碳氮比介乎25-30:1 之間。

不同的原材料涉及不同的碳氮比例，木碎的碳氣比約為400-500:1，菜葉的數值則約為10-15:1。透過調配不同堆肥材料的數量，我們便可得出一個相對合適的碳氮比例。

For microorganisms, both carbon and nitrogen are essential life-supporting elements. Mixing raw materials in a proper C:N ratio can greatly enhance decomposition of organic matters in the composting process. The ideal C:N ratio for composting is 25-30:1.

The C:N ratio differs for different types of materials, for example, the C:N ratio of wood shavings and vegetable leaves are 400-500:1 and 10-15:1 respectively. The C:N ratio of a compost pile can be adjusted by controlling the proportion of different materials.

計算碳氮比例：
Example of calculating C:N ratio:

		碳C	:	氮N
一份樹葉 1 portion of tree leaves		40	:	1
一份水果 1 portion of fruit		30	:	1
一份蔬菜 1 portion of vegetable leaves		10	:	1
總量 Total content		80	:	3
整體碳氮比 The overall C:N ratio		27	:	1

材料表面積
Surface Area

較大的材料表面積，有利微生物更有效地分解材料。可將較大的材料弄成小塊（如將枝條切至5-10cm），這樣有助增大面積，加快堆肥速度。不過，材料弄到太碎也不好，這樣容易令材料過於結實而導致透氣不足。

The larger surface area the microorganisms have to work on, the faster the materials will decompose. By breaking the materials into a smaller size (eg. break twig into 5-10cm pieces), one increases the surface area, and thus speeds up composting. However, avoid breaking the materials too fine or they get compacted and diminish aeration.

溫度
Temperature

堆肥過程中溫度會上升，而微生物在溫暖的環境下亦會較為活躍。堆肥的理想溫度大約為攝氏60度。

以下是一些基本的保溫措施：

- 堆肥箱不要處於當風或撇雨的位置
- 利用帆布等不透水材料作為覆蓋
- 堆肥材料的體積不少於1立方米

Heat is released during the composting process and microorganisms are more active in a warmer environment. The optimum temperature is around 60 degrees.

Here are some ways to prevent heat loss from the compost pile:

- Avoid setting the composting bin at a location where it is exposed to wind or rain
- Cover the compost pile with water-proof materials (eg. plastic sheet or carpet)
- The size of compost pile should be $1m^3$ or above

水份
Water

與人類一樣，微生物亦需要水份維生，材料乾燥時，堆肥的生物作用亦會停止。堆肥材料的理想濕度為70%，如要測試水份含量是否適合，可隨意拿一撮堆肥材料，揉作一團，如材料可握成圓球形狀而又無水滴出，便大致正確。

水份太低或太高都會帶來不良影響，太乾的話微生物活動會受到限制，太濕則會導致透氣不足而缺氧，兩個情況都需要小心預防。如水份不足，可加入適量的水，或一些水份含量較高的物料，如蔬果或濕報紙。至於水份過多，則應改善堆肥箱的去水功能，勤收堆肥液；亦可將堆肥扒開，讓材料暴露於空氣中。

Just like people, microorganisms need water to live. Biological activity stops when the pile dries out. Ideally, the overall moisture content of the composting materials should be 70%. To assess whether the moisture content is sufficient, pick a handful of materials and knead it. If the materials can be shaped into a ball and no water is dropping out of it, the moisture is about right.

Excessively high moisture content should be avoided, as it displaces air from the pores between the soil particles and causes anaerobic conditions – the pile will stink. However, too low moisture content deprives organisms of the water needed for their metabolism, and inhibits their activity.

The moisture content of a compost pile can be raised by adding either water or materials of high water content (such as vegetable, fruit and soaked newspaper). On the contrary, moisture content can be reduced by draining out 'compost-tea', or spreading the compost pile to increase water evaporation.

酸鹼度
pH Value

微鹼至中性(ph7-8)的環境，是最適合堆肥的微生物活動的。由於分解過程中會產生有機酸，因此材料的酸鹼度在後期會出現下降。可透過加入少量的熟石灰或草木灰，讓微生物保持活躍。

Microorganisms are generally more active under a neutral (pH=7) or slightly alkaline (pH7-8) environment. In the initial stages of composting, organic acids are formed and pH drops. By adding a small amount of hydrated lime or plant ash, it neutralizes the compost and keeps the microorganism active.

製作堆肥的步驟！
Procedures for Compost Making!

Step1

準備好堆肥箱。

Buy or make a compost bin

Step2

收集各種堆肥原料，按照合適的碳氮比例，混合一起。

Collect compost materials and mix them in the proper C:N ratio

Step3

從溫度、顏色、形狀、氣味、濕度等不同層面，監察分解的進度，並按實際變化作出調整。

During the composting process, monitor the temperature, colour, odor, moisture content and shape of the compost pile. Make timely interventions for adjustment.

Step4

定時翻堆，每個月翻堆2-3次；整個堆肥過程約需兩至三個月。

Turn the compost two or three times each month for two to three months.

Step5

堆肥材料成熟時，會有以下特徵：
- 成為較細緻的混合物，呈深褐色
- 不帶臭味，氣味反而有點像泥土
- 體積下降至原來的60%左右
- 溫度與外間相仿，並保持平穩

Mature compost will have the following characteristics:
- A mixture of fine substances which are dark-brown in colour
- It does not stink but gives off an odor like soil
- It is reduced to about 60% of the original size
- The temperature is steady and similar to that of the surrounding environment

製作堆肥箱
Compost Bin

設計堆肥箱時，要考慮到通風、耐用性、衛生、容量及製作成本等因素。以下介紹兩款適用於社區農圃的堆肥箱。

Aeration, durability, hygiene, volume and cost are all important factors to consider when choosing a compost bin. Here are the two types of compost bin that are appropriate for community farming.

① 三格開放式堆肥箱

Three-compartment open composter

三格的堆肥箱設計，需要佔用較多的地方，但能處理較大數量的有機廢料，並可分階段處理。當第一格材料已滿、需作翻堆時，可將材料翻堆至第二格，然後按照情況再移至第三格。堆肥箱底部可鋪上一些泥土，以吸收堆肥材料釋放出來的堆肥液。

箱的結構可以用磚、鐵絲網或木板來製造：

- **磚頭堆肥箱**：磚頭之間保留少許空隙以通風，前牆以木板製造，以便開關。
- **鐵網堆肥箱**：以鐵柱或木樁作支架，形成三格，再以鐵網圍繞。
- **木製堆肥箱**：設置三格的支架後，再以長木板圍繞。

The three-bin system occupies a larger space but it has a higher capacity and enables one to have three different piles at various stages of composting. By starting the pile on one end, you can move it to the second compartment when the first one is full and the compost is ready to turn, then repeat, turning compost to the third bin, where it finishes. Lay some soil on the bottom of the bin before operation, so as to hold the liquid released during the composting process.

The system can be built with bricks, wire mesh or plywood:

- If building with brick, leave some space between bricks to facilitate aeration. Cover the front-side with a wooden panel for ease of operation.
- If building with metal mesh, use metal or wood poles to set the framework for the three compartments. Cover the sides with metal mesh.
- If building with plywood, firstly make the frame with wooden poles and then cover the sides with plywood panels.

② 關閉式膠桶堆肥箱
Closed compost bin

適合缺乏空間的家居或農圃，可找一個連蓋、可防蟲的大膠桶作為外箱，垃圾桶或洗衣桶也可以使用。

這種堆肥箱，需要較頻密的翻堆以保持透氣。另外，由於它的容量較少，可能需同持設置數個桶，以便能夠持續地收集有機廢物。

This system is more suitable for a household situation where space is always limited. Source a plastic container with a lid to address the nuisance related to pest and odor. A rubbish bin or a container of laundry power works for this purpose. This bin will require more frequent turning to enhance aeration. More than one bin may be needed for a continuous operation.

製作步驟
Instructions

1. 將鐵網圍繞成圓桶形狀，外圍再繞上黑網。
2. 膠桶外面鑽上數個小孔，以便透氣。
3. 桶的底部設水龍頭位，以收集堆肥液。
4. 將鐵網桶放入大膠桶之內。

1. Make a wire net cylinder of a size that fits into the container.
2. Drill small holes all over the plastic container.
3. Install a tap at the lower end of the plastic container for draining out 'compost tea.'
4. Fit the wire-net cylinder into the plastic container.

常見堆肥Q&A
Frequently Asked Questions

Q1
製作堆肥需時多久？
How long does it take to make compost?

視乎溫度、通風、原材料的類別及體積、濕度和管理方法等。在香港，一個完整的堆肥過程大約需要二、三個月時間。

It depends on various factors such as temperature, aeration, types and sizes of materials, humidity and management. In Hong Kong, it takes at least two to three months to complete one composting process.

Q2
可以使用未完全腐熟的堆肥嗎？
Can I use the compost before it is fully decomposed?

使用未腐熟的堆肥，可能會對農作物造成損害！因為這時材料仍處於分解階段，釋放的熱力會對植物根部造成影響；而材料內的微生物，也可能會與農作物競爭氮肥。

Application of pre-mature compost will result in crop damage. As the decomposition process is still ongoing, the heat it generates will cause root damage. The microorganisms will also compete with the crop for nitrogen in the soil during the decomposition process.

Q3
堆肥已變乾，但仍未完全分解，為甚麼？
Why does my compost dry out before it is fully decomposed?

有兩個可能原因，第一是高碳材料的比例太大，第二是水份過度蒸發。如果是碳材料太多，可考慮適量加入高氮材料如豆渣，以作平衡。如果是水份蒸發太快，則應先補充水份，再以帆布等覆蓋材料，減少水份流失。

Biological activity stops when the compost pile dries out, which may be due to excessive carbon content or high water loss. To maintain a proper moisture content, you may add water to the compost pile, turn it, and cover it with a plastic sheet to prevent water loss. If the carbon content of the compost pile is too high, lower it by adding materials which are rich in nitrogen content such as soy meal.

Q4

怎樣避免蒼蠅？
How can I get rid of flies from the compost bin?

廚餘材料比較容易引來蚊蟲，使用時必需先與木糠、泥土等充份混合，並以帆布等作覆蓋。此外，定時翻堆及添加少許熟石灰，也可以幫助減少蒼蠅。

Food waste tends to attract pests. When you are handling food waste, always mix it with wood shaving or soil before adding it to an open compost bin. Cover the bin with a plastic sheet and turn the compost more regularly to help to control the pests. You may also mix hydrated lime with compost materials to repel pests.

Q5

堆肥材料沒有明顯升溫，為甚麼？
Why doesn't my compost warm up?

可能的原因有幾個，包括氮材料不足、熱量流失太快、水份過多或太少、堆肥材料數量不足等。你可考慮以下方法來提升效率：

• 添加高氮的材料，如菜笑。

• 如材料太乾，補充水份；如太濕，添加較乾的材料；如失溫太快，可以帆布覆蓋保溫，或將堆肥箱遷至一個比較不當風的位置。

• 材料數量應不少於1立方米。如不足，可加添材料再重新開始。

It may be due to low nitrogen content, rapid heat loss, excessive or insufficient moisture content, and the compost pile being too small in size. To enhance the composting process, you can:

• Add materials of high nitrogen content, such as vegetable leaves, to the compost pile;

• Add water if the compost pile is dry, or add wood shavings if the pile is too wet. Cover it with a plastic sheet to preserve moisture content and prevent heat loss. You may also consider relocating the compost area if it is too exposed to wind.

• The compost pile should be $1m^3$ or above in size. Make a bigger pile and restart the process.

堆肥籃
Basket Composting

堆肥籃的設計，是仿傚自然界養份循環的原理，生物體死去或樹葉跌落至地面，都會被鄰近的微生物、真菌及無脊椎動物所分解，轉化成較簡單的形態，被四周的植物所吸收。

在園圃中設置堆肥籃以收集作物的殘株，可讓堆肥變得更方便及直接，節省時間及管理工作。

製作堆肥籃的步驟：

1. 在農圃中設置一個籃子，一個4平方米農圃大約需要一個20公升以上容量的堆肥籃。

2. 籃子應放置在容易收集菜筴、植物殘株的農田上面。

3. 監察材料的分解情況，如太濕，可撥開材料以增加蒸發，有需要時可加上少許熟石灰。

4. 大約兩至三個月後，便能翻起堆肥材料，作為土壤的改良劑。

Basket composting imitates the nutrient cycle in nature, where dead organisms and fallen leaves on soil are decomposed by the surrounding microorganisms, fungi and invertebrates. The decomposition process breaks down organic substances into a simple form to be taken up by the surrounding plants.

By placing a composting basket right in the garden to collect crop residues, it makes composting more convenient and straight forward, which saves time and labour.

Steps for basket composting:

1. Source a basket of compatible size to the garden. A $4m^2$ garden needs a basket of 20 litres or above in size.

2. Place the basket on top of the soil in the garden where it is convenient for dumping crop residue and where it is surrounded by crops.

3. Monitor the compost pile – if it is too wet, turn and spread the content to enhance water evaporation. Add hydrated lime on top if necessary.

4. After two to three months, empty the basket and apply the compost for soil enrichment.

波卡西(發酵)堆肥
Bokashi composting

甚麼是波卡西堆肥？

波卡西(Bokashi)是一個源自日本的詞語，解作發酵的意思。運用波卡西方法製作堆肥，即是透過發酵方法把有機物分解成植物可吸收的養份。

波卡西堆肥的運作方法

- 通常，運用波卡西的人會使用專為此方法設計的波卡西堆肥箱，如果想自己造堆肥箱，要留意堆肥箱的蓋需能夠完全密封整個堆肥箱，並在箱底對上10cm的位置設一個隔篩，下面安裝一個水龍頭收集堆肥液。
- 操作者可以先把廚餘或有機廢料切碎，放進箱內，然後加入波卡西粉將材料覆蓋。重複以上步驟，直至填滿堆肥箱。
- 廚餘在箱內發酵，這是一種厭氧反應，特點是分解得比較慢。由於堆肥箱是密封的，異味不易溢出。這種堆肥方法的缺點是有機物只能分解至半成品，最後也需要埋在泥土裏，讓土壤的微生物完成最後步驟。已埋入波卡西廚餘的泥土，過了2-4星期後，便能開始種植作物。

What is Bokashi composing?

The word "Bokashi" originates from Japan, meaning fermentation. In Bokashi composting, organic matter is degraded to a nutrient-rich liquid through a fermentation process. Plants can easily absorb the nutrients in this form.

Operation of Bokashi composting

- A compost bin that is tailor-made for Bokashi composting is commercially available. You can also make your own compost bin by modifying a bucket that can be tightly covered by a lid. Lay a sieve inside the bucket at 10cm above the bottom, to separate solid and liquid content. Install a tap below the sieve outside the bucket for collecting of liquid compost.
- Collect kitchen waste and garden waste for compost-making. First chop the waste materials into small pieces and put them in the bucket. Apply a layer of Bokashi on top of the waste. Repeat the process if more waste is added to the bucket until it is full.
- When the materials ferment inside the bucket, anaerobic decomposition takes place slowly. Close the lid of the bucket tightly to avoid leakage of the unpleasant smell. Liquid can be collected regularly and diluted for application as fertilizer. In Bokashi composting, organic matter is not fully decomposed inside the bucket. When the bucket is full, bury the semi-decomposed product in soil for soil microbes to act to complete the decomposition process. You may plant into the bokashi enriched soil 2-4 weeks after it has been buried.

把波卡西粉與廚餘混合
Mix the bokashi powder with the food waste

埋下已發酵的廚餘
Bury the fermented food waste

Chapter 04

作物篇 Vegetables and Herbs

	可種植之月份 Planting Month
	適合種植之節氣 Best Sowing Time in Solar Term
	泥層厚度（cm） Soil Depth
	生長日數 Growth Period
	種植距離 Planting Distance

香港位於亞熱帶，每年約有一半時間，氣溫相當和暖；雨量亦充沛，由東平洲約1400毫米至大帽山一帶的3000毫米不等。香港的氣候適合多種不同類型的農作物生長，單是蔬菜，合適品種已達約七十種；連同生果、香草及水生作物的話，種植清單更超越一百之數。

有趣的是，這個作物清單是會隨時間而演變的。上世紀五十年代以前，香港以種植稻米為主，戰後隨着人口及市場的轉變，逐漸變成以蔬菜為主的格局。由於飲食習慣及潮流的轉變，近十數年興起了一些另外的品種，如紅菜頭、秋葵、翠玉瓜、羽衣甘藍等；同時，也有一些品種因市場轉變、乏人問津而沒落，如唐生菜。

農作物的歷史清單，大概還會繼續演變下去。不過，我們應堅守永續農業「不時不植」的原則，適合的品種，我們努力把它種好；不適合的，我們別以非環保手段逆天而生，強行栽種。

本章，我們會介紹七十多種適合香港氣候栽種的農作物，包括蔬菜、雜糧、生果、香草等。它們絕大多數都可以在地方較狹窄的家居環境中種植。

Hong Kong's climate is sub-tropical, tending towards temperate for nearly half of the year. The mean rainfall ranges from 1400 millimeters at Ping Chau to more than 3000 millimetres in the vicinity of Tai Mo Shan. There is a wide range of crops suitable for planting under this climate – more than 100 varieties of crops can be grown in Hong Kong, of which around 70 are vegetable.

Crops produced in Hong Kong have been evolving over time. Rice growing was dominant before the 1950's, and vegetable growing has been gradually taking over our farmland, along with the population growth, after Second World War. Other varieties like beet root, okra, zucchini and Toscana kale become popular in recent decades due to the change in food culture. The market demand for the common vegetable varieties of the old days, like Chinese lettuce, has been fading.

Although market demand for vegetable has always been changing, the rule for sustainable farming, planting in season, stays unchanged. We should farm according to the natural, climatic condition rather than against it.

In this chapter, we will introduce more than 70 types of vegetable, herb, grain and fruit, which are suitable for growing in Hong Kong. Most of them can also be planted in confined spaces at home.

葉菜類
Leaf Vegetables

蔬菜之中，葉菜可說是最重要的一個類別。在華南地區，秋播的葉菜款式種類較多，春播明顯較少。由於消費者對葉菜的需求甚殷，近年也出現了一些不太值得鼓勵的種植方法，例如部份生產者不惜夏季在中國北方地區種植葉菜，然後長程運送抵香港，以求本地夏季亦有相關的葉菜供應。這方法操作上雖然可行，但卻涉及大量的能源消耗。

以永續農業的角度分析，我們還是建議消費者「不時不食」；至於作為蔬菜種植者，也應以「不時不植」為原則。

除了人類，雀鳥也是葉菜類蔬菜的主要「消費者」，需要預防鳥害。除了懸掛光碟、反光紙、彩色氣球等方法，也可以選擇網眼不大於半吋、不會傷及雀鳥的防鳥網，或佈置塑膠蛇、猛禽模型在農田之中，用來保護農作物。

Leaf vegetables dominate the vegetable category. In southern China, varieties of leaf vegetable that are suitable for sowing in autumn are more diverse than that of spring. To cope with the local demand for leaf vegetables in summer, traders transport produce from Northern China to Hong Kong at the expense of a high energy footprint. This food production and supply chain is not sustainable.

From a sustainability perspective, farmers should produce food according to the seasonal rhythm, while consumers should also only buy and consume food in season.

Wild birds enjoy leaf vegetables as much as humans. To live in harmony with wildlife, we could use old CDs, reflective paper and colorful balloons as deterrents or a net with a mesh-size smaller than 1cm for crop protection. Rubber snakes from a toy shop and even plastic birds of prey can also be effective.

田畦建議
Soil Depth

田型的佈置，對農作物根部的發展有決定性的影響。因此，介紹種植方法時會對田畦厚度作出具體建議（淺畦 10-15cm；中畦 15-20cm；高畦 20cm 以上）。如果採用盆栽種植，請將田畦厚度對換成盆栽中的泥土厚度。

Soil depth is critical for healthy root growth. Soil depth requirements of different crops are broadly defined as shallow (10-15cm); medium (15-20cm) or deep (20cm or above) in this chapter. When planting in a container, choose one that suits the plant's soil depth requirement.

茄科
Solanaceae

葉菜
Leaf Vegetable

淺田畦
Shallow

高田畦
Deep

白菜 White Cabbage (Pak Choi)

1. 上海白菜
   Cabbage Shanghai
2. 高腳白菜
   Long-petiole dark-leaf Pak Choi

白菜是一個很籠統的稱呼，當中包括許多不同的品種，例如高腳黑葉、中腳黑葉、矮腳黑葉、春水白、鶴藪白和上海白等。白菜主要是秋播的農作物，但不同品種之間的抗耐性有很大差別，鶴藪白會較難栽種，高腳品種則較能忍耐濕熱氣候。種植時，需保持田畦平坦，及勤澆水。

White cabbage is a general name for a range of varieties covering Long-petiole dark-leaf pak choi, Middle-petiole dark leaf pak choi, Short-petiole dark leaf pak choi, Chunshui pak choi, Hoktau pak choi and Shanghai cabbage. The tolerance to growing conditions varies among varieties. Hoktau pak choi is relatively less tolerant, while the long-petiole varieties are more resistant to hot and humid weather. White cabbage is suitable for sowing in autumn. Level soil and frequent watering are required.

 4月-10月
April to October

 處暑
End of Heat

 淺(10-15)
Shallow

40-70日
40-70 Day

10-20厘米
10-20 cm

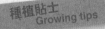
種植貼士
Growing tips

- 忌旱。
- 容易受鳥害，需小心預防。
- 12月開始，狗蝨仔的侵害會變得嚴重，應避免遲種。
- 開花前收採，以免莖葉變得太老而不堪食用。

- Low tolerance to dryness.
- Protection against bird damage is necessary.
- Avoid planting in December when flea beetle (*Phyllotreta striolata*) is getting active.
- Harvest before flowering, otherwise leaves are too old to be eaten.

菜心 Flowering Cabbage

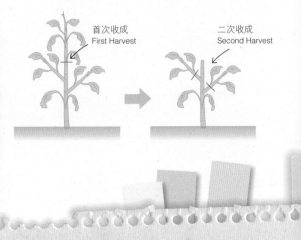

首次收成
First Harvest

二次收成
Second Harvest

菜心的品種以開花日數來區別,50日、60日、70日菜心均有。一般來說,日數較少的菜心,菜薹會較幼,也會較為耐熱。俗稱「四九菜心」的短日數菜心,甚至可以在夏季種植。

菜心的收成時機,是在「抽薹」、剛剛開花的時候,這時菜薹的品質最為幼嫩。收割菜心的日子必需準確,太早菜薹尚未成型,太遲則會老化。收採主薹後,植株會在收割位下面的節位長出側枝,可作為第二次收採。情況良好的話,一棵菜心可以收成三至四次。

田間管理方面,方法與白菜大致類同。

Flowering cabbage is classified according to the number of days to flowering, as 50-, 60- and 70-day varieties. Generally, the shorter the number of days to flowering, the more heat-resistant the variety is. A variety named 49-flowering-cabbage can be grown even in summer. The best time for harvesting is when the plant starts flowering. At this time the flowering stalk is tender and tasty. Harvest timing has to be precise. Further harvest of lateral branches is possible and up to four harvests could be made in the whole growing period. The daily management is similar to that of white cabbage.

4月-10月
April to October

處暑(四九菜心可較早播種)
End of Heat
(49-flowering-cabbage
can be sowed earlier)

淺(10-15)
Shallow

40-70日
40-70 Days

8-20厘米
8-20 cm

種植貼士
Growing tips

- 12月開始,狗蚤仔的侵害會變得嚴重,因此應避免遲種
- 剛開始生長菜薹時收採

- Avoid planting in December when flea beetle (*Phyllotreta striolata*) is getting active.
- Harvest when it starts flowering.

100

芥菜 Mustard Leaf

芥菜的品種類別，可能較白菜還要多。既有體積細小的南風芥，也有植株高逾尺、可作菜乾的竹芥、大芥菜(見圖)，此外還有包心芥菜，以及用作泡製醃菜的根用芥菜。

由於芥菜帶苦味，鮮吃方面不及白菜普遍，但許多時可用來製作菜乾或醃菜，提供另外的選擇。種植方面，較大品種之芥菜，所需的管理密度及灌溉次數，都會較低。

Mustard leaf has wider varieties than pak choi. Well known varieties include Nanfeng mustard, which is small in size, Zhu mustard and large mustard (pictured) of 30cm in height which are often processed to dried vegetables, and head mustard and root mustard that are used for pickling. Mustard leaf's bitter taste makes it less popular than pak choi for direct consumption. Large-sized mustard requires less frequent watering and management.

4月-10月(視乎品種) April to October (depends on variety)	
處暑(南風芥可較早播種) End of Heat (Nanfeng mustard can be sowed earlier)	
淺至中(10-20，視乎品種) Shallow to Medium (depends on variety)	
40-80日 40-80 Day	
10-40厘米 10-40 cm	

種植貼士 Growing tips

- 12月開始，狗蚤仔的侵害會變得嚴重
- 開花前收採

- Damage by flea beetle (*Phyllotreta striolata*) is severe in December.
- Harvest before flowering.

芥蘭 Chinese Kale

收採主薹(折口處)後，芥蘭還可以續收側枝。
After the main stalk is harvested, further harvest of lateral branches can be made.

芥蘭是冬季的主要蔬菜。由於其葉片較厚，受狗蚤仔的影響較為輕微，因此收成期會比白菜及菜心長；芥蘭的病害也較菜心少，是一個產量較穩定的作物。

芥蘭也有大、小之分，現時市面較常見的是中花芥蘭，此外也有專吃主薹的荷塘芥蘭，以及較耐熱的玉芥蘭品種。與菜心一樣，芥蘭都是收採剛「抽薹」的菜薹，可多次收採。在生長階段中期需追施氮肥，為較後時期的側枝提供養份。

Chinese kale is a major crop in winter. Because of its thick leaf, it is more resistant to disease and damage by flea beetle. It has a longer planting period and gives a more stable yield than pak choi and flowering cabbage. Chinese kale has large and small-sized varieties. Well known varieties include the commonly seen Middle maturity Chinese kale, Hetang Kale that is well-known for its tasty flowering stalk and the heat-resisting Jade Chinese Kale. In the middle of growing stage, apply nitrogenous (e.g. peanut cake) fertilizer to provide nutrient to lateral branches.

8月下旬-12月
Late August to December

處暑(中花)；
秋分(荷塘)
End of Heat (Middle maturity Chinese Kale); Autumn Equinox (Hetang Kale)

淺至中(10-20，視乎品種)
Shallow to Medium (depends on variety)

60-80日
60-80 Days

15-25厘米
15-25 cm

種植貼士 Growing tips

- 與菜心一樣收採菜薹，由於枝條較為粗壯，一般收成會較菜心為多

- Harvest when it starts flowering.Lateral branch growth is strong, which produces higher yield than flowering cabbage.

椰菜 Cabbage

1. 青椰菜
   Green Cabbage
2. 紅椰菜
   Red Cabbage

椰菜也是華南地區的重要蔬菜。它的管理工作遠較白菜、菜心為少，種植起來較為輕鬆。不過，雀鳥及菜青蟲是椰菜的大敵，要小心留意及預防。

椰菜有早水、遲水、尖頂等品種，也有主要用作沙律、西湯或伴碟的紅椰菜，栽種的方法大同小異。較特別的品種，是收採包心側芽的「抱子甘藍」(俗稱椰菜仔)，但這品種要在較涼的氣候才生長得好，在香港種植沒有太大把握。此外，也有不少花葉、雜色的觀賞椰菜品種，但較少用作煮食用途。

椰菜定植後，建議在田畦面上鋪設護根，這樣可有效減少生長中、後期的澆水次數。

Cabbage is another major vegetable crop in Southern China. It requires less daily management than pak choi and flowering cabbage but proper protection against damage by birds and caterpillars is essential. Common cabbage varieties include early-planting, late-planting and pointed-top, as well as red cabbage which is commonly served in salads, soups and side dishes. The planting method is similar for all species. Brussel sprout is a special variety where headed lateral buds are harvested. It prefers cold weather and is not suitable for growing in Hong Kong. There are also some ornamental varieties. Full mulching is recommended after transplanting, and there is no need for frequent watering in the middle and late growing period.

 8月下旬-12月
Late August to December

 早水品種可在處暑種植，約20天後可種遲水、紅椰菜等品種
Sow early planting varieties after 'End of Heat' and late- planting varieties such as red cabbage three weeks later.

中至深(15-25，視乎品種)
Medium to Deep (depends on variety)

 90-120 日
90-120 Days

 40-50 厘米
40-50 cm

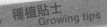
種植貼士
Growing tips

- 適合以免掘方法種植
- 及早提防及清除菜青蟲
- 椰菜充份包心、顯得結實時，便能收採

- Cabbage is suitable for growing with no-dig method.
- Beware of caterpillars of Small Cabbage White and immediate removal is essential for prevention of pest outbreak.
- Harvest when the round-shape leaf head becomes sturdy.

黃芽白 Chinese Cabbage

8月下旬-9月
End of August to September

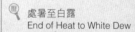
處暑至白露
End of Heat to White Dew

中（15-20）
Medium

70-90日
70-90 Day

30-40厘米
30-40 cm

黃芽白及紹菜等大白菜品種，本屬北方蔬菜，在華南地區只能在秋冬時栽種，由於種植條件不同，所以也沒有太大的把握。

Chinese cabbage and Tientsin cabbage originate from northern China. They can only be planted during winter in Southern China.

種植貼士 Growing tips

- 低溫有利黃芽白包心，惟香港冬季往往不夠冷
- 12月開始，狗蝨仔的侵害會變得嚴重
- 充份包心、顯得結實時，便能收採

- The dense-leaved heads form occurs in low temperature but Hong Kong's winter is usually too warm.
- Damage by flea beetle (*Phyllotreta striolata*) is severe in December.
- Harvest when the leaf head becomes sturdy.

生菜 Lettuce

生菜是秋冬季節最常種植的蔬菜。許多年前，生菜以唐生菜及西生菜(包心生菜)兩類為主。但自意大利生菜出現後，唐生菜已被取代。此外，農夫亦從外地引進了許多不同品種的生菜，例如牛油生菜、羅馬生菜、沙律生菜等等。生菜的管理工作雖然不算少，但種植難度不高，病蟲害亦少。只要願意投放時間與精神，不難取得好收成。

生菜需及時澆水，以保持泥土濕潤，否則菜葉會易老及帶苦味。除了一次性的收採方法外，部份園藝種植者會選擇同時種植多款不同的生菜，每次收成時，每棵摘下兩、三片葉(從外圍收採起)，其餘部份繼續生長。這樣的話，生菜的收成期可持續至一個多月。

Lettuce is the most commonly grown produce in winter. Chinese lettuce and head lettuce used to be the major varieties, but Italian lettuce has replaced the Chinese lettuce in recent years. Local farmers have also introduced other varieties from foreign regions, such as butter lettuce, Romaine lettuce and different varieties of salad lettuce. Lettuce is easy to grow and less vulnerable to pest and diseases. Planting lettuce offers a rewarding experience.

Frequent watering is needed or the taste of produce will become bitter. Rather than harvesting the whole lettuce, the home-grower can harvest a few peripheral leaves from several plants each time to extend the harvesting period.

 8月下旬-2月
Late August to February

 處暑
End of Heat

淺(10-15)
Shallow

60-70日
60-70 Day

20-25厘米
20-25 cm

種植貼士 Growing tips

- 水份不足會引致苦味，宜及時澆水及鋪設護根
- 意大利生菜較耐熱，可收採至六月
- 收成：

  中式：充份成長後，作一次性收採

  沙律：每棵每次收2至3塊葉

- Insufficient irrigation results in bitter-tasted lettuce. Regular irrigation is needed and mulching helps to reduce water loss.
- Italian lettuce is more tolerant to heat and its harvesting period is June.
- Harvest:

  For Chinese Dish: One off harvest when the lettuce become full size.

  For Salad: Pluck off 2-3 leaves per lettuce.

油麥菜 Indian Lettuce

油麥菜可算是生菜的一個分支。

在北方，也有種植主要收採菜薹的品種，稱為萵筍，但在香港較為少見。

Indian lettuce is a sub-category of lettuce. Asparagus lettuce is a less common variety grown in northern China, of which the flowering stalk is harvested.

8月下旬-2月
Late August to February

處暑
End of Heat

淺(10-15)
Shallow

70-80 日
70-80 Day

25-30 厘米
25-30 cm

種植貼士
Growing tips

- 種植方法與牛菜類同，但所需空間較多
- 充份成長後，作一次性的收採

- Planting method is similar to that of lettuce, except more space is needed.
- One off harvest when Indian lettuce becomes full szie.

茼蒿 Garland Chrysanthemum

茼蒿花
Flower of Garland Chrysanthemum

茼蒿與生菜、油麥一樣，都是菊科的農作物，種植條件亦相類似。相較下，茼蒿的病蟲害較少，更容易種植。

由於茼蒿的強烈味道，並非所有人都能接受，所以，近年本地也開始引進一些味道較清淡的茼蒿品種，例如春菊（日本茼蒿）及皇帝菜。

Lettuce, Indian lettuce and garland chrysanthemum are all vegetables of the Asteraceae family and their growing requirement is similar. Garland chrysanthemum is more pest-resistant and therefore easier to grow. Garland chrysanthemum is strong in flavor, but milder varieties like Japanese garland chrysanthemum and 'Emperor Vegetable' have been introduced in recent years.

8 月下旬 -10 月
Late August to October

處暑
End of Heat

淺 (10-15)
Shallow

50-60 日
50-60 Days

15-20 厘米
15-20 cm

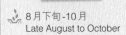

種植貼士
Growing tips

- 可連續收成，每次收採枝條頂部 10-15 厘米的嫩葉部份，種植中期需施氮肥（如花生麩）補充養份
- 需及時澆水，否則會導致茼蒿帶苦味
- 翌年 3-4 月開花，具觀賞價值；早播有利延長收成期

- Continuous harvest. Crop the top 10-15cm of young shoot. Fertilizer application is required during planting period.
- Insufficient irrigation results in bitter-tasted crop.
- Flowering in March to April, with ornamental value. Early sowing allows longer harvesting period.

菠菜 Spinach

冬季主流蔬菜之一。菠菜本身病蟲害不多，種植方面不算困難，但管理上會較花功夫，包括疏苗、除草、澆水等等，種植者宜先作衡量。

Spinach is a popular winter crop. It is not prone to pest and disease, but requires relatively high management input throughout the planting process, including thinning, weeding and watering.

9月-2月
September to February

白露
White Dew

淺（10-15）
Shallow

60-70日
60-70 Day

15-20厘米
15-20 cm

種植貼士 Growing tips

- 菠菜喜冷，收成期至翌年4、5月結束
- 菠菜根系的再生能力很差，忌移植，宜露地直播，可分批播種以延長收成時期
- 及時疏苗、除草、澆水及追肥
- 抽薹前收成，收採時需整株拔起

- Spinach likes cold weather. Harvesting ends by May.
- Weak in root regeneration ability. Direct sowing. Do not transplant seedlings. Sow different batches to prolong harvesting.
- Thin and weed regularly. Requires frequent irrigation and top-application.
- Harvest before flowering stalk grows; Harvest by uprooting the whole plant.

君達菜 Swiss Chard

本地的君達菜，又稱為豬𦡑菜，高逾一呎，呈青綠色。看見這個名字，大抵也會估到它是一種比較粗糙的蔬菜。在外國也有類近品種，植株較本地君達菜更大，並有黃、橙、紅等鮮艷顏色（見圖）。君達菜的管理工作不多，可說是一種「懶人蔬菜」。

Swiss chard is nicknamed "piggy vegetable" and 'vegetable for the lazy one' locally because it is so easy to grow and requires little caring. The local variety is green in colour and grows up to 30cm tall. Varieties of yellow, orange and red are available.

🌱	9月-10月 September to October
🔍	白露 White Dew
🌱	中（15-20） Medium
🌿	70-90日 70-90 Day
🌱	40-50厘米 40-50 cm

種植貼士 Growing tips

- 根系再生能力差，宜採用帶穴苗盤來育苗；移植時避免根部受損
- 可採免掘、全覆蓋等方法，保持泥土濕度
- 可連續收成，至翌年5月。每次收成時每株只會摘去二、三塊葉片，由外圍開始收採，生長中段要追施1、2次肥料。早播種有利延長收成期。

- Propagate in tray with cells. Transplant seedlings with care to avoid root damage.
- Apply no-dig method and mulching after transplanting, to retain moisture.
- Continuous harvest is possible until May. Harvest the outer leaves. Apply fertilizer 1-2 times in the middle of growing period. Early sowing allows a longer harvesting period.

芹菜 Celery

芹菜主要分唐芹及西芹（見圖）兩類，唐芹的植株較小，枝段較幼，但味道較香濃，常用於煮湯或配料；西芹的味較淡，但枝段較粗，可作主食。兩者的種植方法大同小異。

芹菜的生長相對緩慢，苗期也要接近兩個月，9月播種，翌年2、3月才可收成，差不多要花上半年時間。不過，種植期間芹菜的管理工作並不多，病蟲害亦少。

Chinese and European celery (pictured) are the major varieties. The former is smaller in size with stronger flavor and it is used in soup and cooking for flavoring. European celery has a mild favor and is consumed as a staple vegetable.

Celery grows slowly. It takes two months for the seedling to develop, and the whole growing period involves about half a year. It is not prone to disease or pest and requires little management input.

9月-10月
September to October

白露
White Dew

中 (15-20)
Medium

120-150日
120-150 Day

30-40厘米
30-40 cm

種植貼士 Growing tips

- 根系再生能力差，宜採用帶穴苗盤來育苗；移植時避免根部受損

- 適合免掘栽種

- 西芹莖部如採光太多，會促使其纖維化，吃起來較為「多渣」；所以收成前二、三個星期，可以報紙包圍着西芹中間枝段部份，再以繩綑着，目的是減少莖部接觸陽光。至於唐芹，由於主要用作配料，多採光香味反而更濃，所以無需綑紮

- 莖部枝條顯得紮實時，便能收採

- Propagate in tray with cells. Transplant seedlings with care to avoid root damage.

- No-dig method is applicable in celery growing.

- Wrap European celery stem with paper a few weeks before harvest to yield stems of softer texture. For Chinese celery, if a stronger flavor is preferred, omit the paper wrapping work.

- Harvest when each stem becomes strong and sturdy.

韭菜 Chinese Chive

韭菜的種植方法容易，打理工夫不多，又沒甚麼病蟲害，很適合初學者種植。

現時，市面常見的多是窄葉品種，其實，韭菜也有闊葉品種，此外還有韭菜花。至於韭黃，則是採用遮光、套桶子等方法，不讓葉片接觸陽光，讓其品質較為幼嫩（原理與絪紮西芹一樣）。

Chinese chive is a heat-resistant perennial and is not prone to pest or disease damage. It is suitable for beginners.

Narrow-leaf varieties are far more common locally than the broad-leaf or flowering ones. To grow hotbed chive, cover it to avoid direct exposure to the sun, so that the crop stays soft.

3-4月、9月 March to April, September	
驚蟄、白露 Insects Waken (Spring); White Dew (Autumn)	
淺（10-15） Shallow	
50-70日 50-70 Days	
20-30厘米 20-30 cm	

種植貼士 Growing tips

- 3-4月及9月，都適合分株移植
- 天氣和暖時，韭菜的生長相當迅速，可供持續收採，每次收採約15厘米長度，生長中期需追施肥料
- 適合免掘種植方法

- March to April and September are ideal periods for propagating Chinese Chive by division.
- Fast growing in warm months. Harvest when leaves reach 15cm tall. Continuous harvest. Apply fertilizer regularly.
- No-dig method is appropriate.

莧菜 Amaranth, Chinese Spinach

1. 紅莧菜
Chinese Spinach (Red)
2. 青莧菜
Chinese Spinach (Green)

莧菜是典型「易學難精」的作物。它發芽容易、生長迅速，表面上好像不難種植。其實，莧菜對水肥的要求非常嚴格，管理乏善的話，會變得粗糙多渣、食用價值下降。種植時，田畦必需平坦，亦要有及時、適量的澆水。

莧菜主要分紅莧、青莧兩種。以葉型分類的話，則可分為尖葉及圓葉，但無論是那一種類，種植及管理方法均大同小異。

Growing Amaranth is easy to learn but difficult to master. It germinates easily and grows fast. However, it gets rough in texture if irrigation and soil fertility are not managed properly. Therefore, a level soil and frequent but mild watering are very important.

Red and green Amaranth are major varieties. It can be classified into round- and acute-leaf by shape. The growing requirements are similar for all Amaranth varieties.

🌱	3月-9月 March to September
🔍	驚蟄後 Insects Waken
🌾	淺(10-15) Shallow
🌿	30-40日 30-40 Day
🌱	5-8厘米 5-8 cm

種植貼士 Growing tips

- 宜露地播種
- 由於莧菜太小太密，難以在株植之間下肥，可考慮採用液肥，或在澆水前薄施麩粉
- 為免品質變差，當植株仍嫩、未太擠迫時便要及早疏苗
- 六、七月時可考慮在棚底栽種，以避過暴雨

- Broadcast sowing is recommended.
- As seedlings are small and dense, either use liquid fertilizer or water the seedlings after top-dressing with peanut cake.
- Thinning before leaves are too crowded and harvest when the leaves are still young and soft.
- Plant under transparent shelter during June and July for protection against rainstorms.

通菜 Water Spinach

夏季的主要蔬菜。坊間一般以水蕹（又稱白骨）、旱蕹（又稱青骨）來分類。其實，這個名稱有些誤導，因為水蕹其實也可以在旱田上種植，而旱蕹實質上也能夠在水池生長。但無論哪種種植方法，通菜依然是耗水量較大的農作物。

Water spinach is a major summer crop. Green water spinach and white water spinach are the main varieties. Both varieties can be grown in water and they have a high water demand.

3-4 月
March to April

驚蟄
Insects Waken

淺（10-15）
Shallow

50-60 日
50-60 Day

30 厘米（15-20 厘米株距）
30 cm (15-20 cm between plants)

種植貼士
Growing tips

- 可持續收成至 8 月。天氣溫暖時，通菜的新梢生長得很快，集中收割枝條前端約 20 厘米較嫩部份，每隔 2、3 星期可收採一次
- 可露地行播
- 通菜是持續收採的作物，中段需補充肥料

- Continuous harvesting until August. Fast growing in warm days. Crop the 20cm youth shoot from the top. Harvest every two to three weeks.
- Sow in line directly in field.
- Irrigate and apply fertilizer regularly.

潺菜 Basella

非常容易栽種的品種，易打理，病蟲害不多，很適合初學者種植。潺菜本身是攀援植物，但因收採部份是嫩梢，故無需搭棚(除非是留種)。

潺菜的管理簡單，需注意的是水肥的供應。

Basella is not prone to pest or disease. It is easy to grow and suitable for beginners. Basella is a climber but no trellis is needed as only young stems are harvested (except for seed saving). Managing basella is fairly simple, irrigation and fertilizer application are all it needs.

 3-4 月
March to April

驚蟄
Insects Waken

淺(10-15)
Shallow

50-60 日
50-60 Days

30厘米
(15-20厘米株距)
30cm
(15-20 cm between plants)

種植貼士 Growing tips

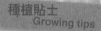

- 天氣溫暖時，潺菜的新梢生長得很快，集中收割枝條前端約15-20厘米較嫩部份，每隔2、3星期可收採一次
- 可露地行播
- 及時澆水及追肥

- Grows fast on warm days. Continuous harvest until August. Crop the top 15-20cm of young shoot. Harvest every two to three weeks.
- Sow directly in the field.
- Regular irrigation and fertilizer application

114

瓜類
Gourds

葫蘆科農作物的品種雖多，但多數具近似習性，包括春播、一年生、抽蔓（需棚架）、耗肥、耗水、枝葉繁茂等。當然，其中也有例外，例如佛手瓜屬多年生、翠玉瓜屬叢生及秋播等。葫蘆科作物的植株較大，除青瓜、苦瓜外，其他品種均需採高畦（20厘米以上），較難採用盆栽來種植。此外，葫蘆科屬異花授粉，不宜單株栽種，最少也要同時種植數株。

Most Cucurbitaceae crops share similar growing characteristics, for instance, spring-sowing, annual in nature, develop lateral shoots, high fertilizer and water demand and they grow rapidly. Nevertheless, there are always exceptions, like chayote is perennial, zucchini does not develop lateral shoots and it is more often being sown in autumn. Cucurbitaceae crops are generally large in size and require a soil depth above 20cm and, therefore, planting in pots is not suitable. As cucurbitaceae crops are cross-pollinated, there is a need to plant several plants together at one time to enhance pollination.

育苗：葫蘆科作物的植株大，生長期長，宜先採用育苗方法，待幼苗具四、五塊真葉時才定植田中。

施肥及護根：定植前，可先以骨粉、堆肥作基肥，定植後追施氮肥。田畦表面建議以護根覆蓋，能保水、防雜草。採用免掘方法來種植瓜類，效果亦不錯。

蟲害：防治瓜蠅是種瓜的一個核心工作。其實，瓜蠅不會吃瓜，只是在果實產卵；皮越薄的瓜，越容易受影響。如果以容易受瓜蠅侵害的程度來排序的話，名單大致如下：
（容易受侵害）↑ 苦瓜、青瓜、節瓜、葫蘆瓜、絲瓜、水瓜、南瓜、冬瓜 ↓（不容易受侵害）
現時，坊間有一些性誘劑，可以用來防治瓜蠅。不過，套網袋仍是最有把握的方法（最適合套袋的時機，是雌花完成授粉、子房開始膨脹時進行，太早套袋會妨礙授粉，太遲則可能已受瓜蠅侵害）。此外，種植者亦應保持良好習慣，及時清理田野中的爛果。

Nursery: A nursery is recommended in view of the large crop size and long growing period. Seedlings should be transplanted when four to five true leaves are developed.

Fertilizer application and mulching: Bone meal and compost can be applied as base fertilizers and topped-up with nitrogenous fertilizer after seedling transplantion. Mulching helps to retain moisture and suppress weeds. The No-dig method goes well with Cucurbitaceae.

Pest damage: Melon fly lay eggs inside gourds and those with a thin rind are most vulnerable to pest damage. Protecting the crop from melon fly damage is essential. The vulnerability of different crops to melon fly damage, in descending order, is Bitter cucumber, Cucumber, Hairy gourd, Bottle gourd, Silky gourd, Sponge gourd, Pumpkin and Wax gourd.
Pheromone-traps are available in the market for melon fly control. However, wrapping each fruit with a net-bag is a more effective measure. The best time for wrapping is when the ovary swells after pollination.
Wrapping too early hinders pollination while wrapping too late risks pest damage. Timely removal of rotten fruit is another good practice for pest control.

青瓜 Cucumber

1, 2. 青瓜 Green cucumber
3. 老黃瓜 Yellow cucumber

青瓜容易種植，可生吃；比起其他瓜類，所佔空間較小，所以很受城市農夫鍾愛。現時市面的青瓜品種非常多，不同形狀及顏色均有。

青瓜還有一個近親品種，就是老黃瓜。兩者種植條件相差不遠，最大分別是青瓜需嫩收，宜勤收採；老黃瓜則要待瓜身充份成熟才能採摘，所以生長期較長。

Cucumber is a popular crop for city farming as it is easy to grow, occupies a relatively small space and the fruits can be consumed raw. There is a wide variety of cucumber of different shapes and colors available in the market.

Yellow cucumber is a kindred variety of cucumber and they share similar growing requirements. The only difference is that green cucumber should be harvested regularly when it is young while yellow cucumber should be harvested when it is fully mature, after a long growing period.

🌱	2 月下旬 - 8 月 Late February to August
💧	雨水 Spring Shower
🌱	中 (15-20) Medium
🌿	50-70 日 50-70 Day
⚃	40-50 厘米 40-50 cm

種植貼士 Growing tips

- 可分批種植以延長收成期
- 當果實仍是深綠色時，便要及時收採
- 需 6-8 呎高的棚子，讓瓜蔓生長
- 露地種植可無需修枝；如以盆栽種植，可酌量修枝以集中養份

- Extend harvesting period by planting several batches of seedlings at different time.
- Harvest when the fruit is still dark green in colour.
- A 6-8 feet high trellis is needed to support growth of lateral shoots.
- Pruning is not necessary for field planting. For pot-planting, prune some of the lateral shoots to preserve nutrients.

苦瓜 Bitter Cucumber

白苦瓜
White bitter cucumber

葫蘆科之中，苦瓜算是較難栽種的品種，主要原因是苦瓜的皮很薄，極易受瓜蠅侵害。可以説，不套袋保護果實的話，種苦瓜基本上是很難有收成的。

除了一般綠色品種，現時市面上也有售價較貴的白苦瓜，味道較清，苦味較淡。白苦瓜結果時需以遮光材料套袋，避免果實受陽光直接照射。

Bitter cucumber is a challenging crop to grow as its thin rind makes it very vulnerable to melon fly damage. Protecting the fruit by wrapping with a net bag is essential.
White bitter cucumber has a milder flavor than the green variety. For planting white bitter cucumber, wrap the fruit with a black plastic bag to prevent sun exposure when it starts to ripen.

2月下旬 - 5月	Late February to May
雨水	Spring Shower
中（15-20）	Medium
60-70日	60-70 Day
45-60厘米	45-60 cm

種植貼士
Growing tips

- 種植方法與青瓜類同
- 當果實仍是光滑及深綠色時，便要及時收採

- Similar to that of cucumber.
- Harvest when the fruit is still smooth and dark green in colour.

絲瓜與水瓜 Silky Gourd and Water Gourd

1. 絲瓜
   Silky Gourd
2. 水瓜絡
   Luffa Sponge

絲瓜（Silky Gourd）

2月下旬-5月
Late February to May

雨水
Spring Shower

深（>20）
Deep

60-70日
60-70 Day

單行種植株距60厘米
Single row.
60 cm between plants.

水瓜（Water Gourd）

2月下旬-4月
Late February to April

雨水
Spring Shower

深（>20）
Deep

60-70日
60-70 Day

單行種植株距60-90厘米
Single row.
60-90 cm between plants.

兩者是近親品種，可雜交，如果留種的話，兩者不能種在一起。絲瓜與水瓜的品種類別相對較少，瓜蠅問題不似苦瓜、青瓜般嚴重。

絲瓜、水瓜還有一個特點，就是可以保留「瓜絡」。當果實老身、乾燥時，果肉會成為纖維網狀，種子藏於其中。這些絲瓜絡、水瓜絡（小圖），可以用作清潔、甚至美容用途。

Silky gourd is a kindred variety of water gourd. As the two varieties cross-breed with each other, they should not be planted closely if seed-saving is expected. Compared to cucumber and bitter cucumber, silky gourd and water gourd are less diverse in variety and less prone to melon fly damage. Aged and dried silky gourd and water gourd are used as luffa sponge. Their fibrous texture makes them good for cleansing and beauty use.

種植貼士 Growing tips

絲瓜
- 早期生長較緩，但後段亦枝葉繁茂
- 需8呎高的棚子，讓瓜蔓生長
- 保留2-3側蔓，其餘剪除
- 當果實仍是軟身、具彈性時，便要及時收成

水瓜
- 需要較大棚子，讓其瓜蔓生長
- 保留2-3側蔓，其餘剪除
- 當果實仍是軟身、具彈性時，便要及時收成

Silky Gourd
- Grows slowly in the early stage but thrives when it is established
- Need 8 feet tall trellis for planting
- Keep 2-3 lateral shoots and prune the rest.
- Harvest the fruit when it is tender and springy.

Water Gourd
- Requires a sun shelter to support its extensive growth
- Keep 2-3 lateral shoots and prune the rest.
- Harvest the fruit when it is tender and springy.

節瓜 Hairy Gourd

節瓜的花
Flower of Hairy Gourd

節瓜的產量豐富，生長迅速，是夏季瓜類收成的主力成員。
節瓜與冬瓜為近親，種植條件亦相類近。

Hairy gourd is a kindred variety of wax gourd and it is one of the major
crops in summer. Fast growing and productive.

2月下旬 - 5月
Late February to May

雨水
Spring Shower

深(>20)
Deep

60-70日
60-70 Day

60厘米
60 cm

種植貼士 Growing tips

- 搭棚、管理方法與絲瓜相類同
- 保留2-3側蔓，其餘剪除
- 當果實直徑接近10厘米時，便要及時收採

- Trellis and other farming requirements are similar to that of silky gourd.
- Remove excessive lateral shoots and leave 2-3 branches to grow.
- Harvest the fruit when its diameter reaches 10 cm.

冬瓜 Wax Gourd

1. 青皮冬
   Wax Gourd (Green)
2. 瓜蔓節位萌發「不定根」
   Adventitions roots
3. 白皮冬
   Wax Gourd (White)

冬瓜是葫蘆科之中，果實最重最大的，一個成熟冬瓜的重量可以超過20公斤。

冬瓜也是屬於「老收」的類別，種植時可順帶留種。但要注意，冬瓜與節瓜屬近親，可雜交，留種時兩者不能種在一起。

Wax gourd's fruit is the largest among Cucurbitaceae crops, weighing up to 20kg.

Wax gourd is often harvested when it is fully mature and therefore seed saving can be done at the same time. As wax gourd is a kindred variety of hairy gourd, cross-breeding is possible. They should not be planted close to each other if seed saving is planned.

 2月下旬-4月
Late February to April

 雨水
Spring Shower

 深(>20)
Deep

90-120 日
90-120 Day

60-90 厘米
60-90 cm

種植貼士
Growing tips

青皮冬：

- 幼苗開始抽蔓時，先將瓜蔓在田面繞一個圈，稱為"盤瓜"，盤瓜的作用是讓瓜蔓的節位萌發不定根，增加植株的根系容量。如果想加強盤瓜的效果，可在盤瓜範圍的節位蓋上一些泥土

- 果實沉重，棚子需堅固

- 只保留主蔓及一個果實

- 果實充份成熟，枝葉開始呈枯黃時才收成

白皮冬：

- 較粗放管理，瓜蔓爬地而生，不需上棚

- 無需修剪，可在節位蓋上泥土，以刺激不定根生長

- 結果時表面會長出一層白粉，這層白粉可以幫助保護果實。不過，為了減低損耗，當白皮冬開始坐果時，最好還是在其底部墊上一些乾草

- 果實充份成熟，枝葉開始呈枯黃時才收成

Green-skin wax gourd

- When lateral shoots emerge, run the shoots on the ground in a circular loop. Cover the loop with soil to encourage growth of adventitious roots.

- It needs a strong trellis to support its heavy fruit.

- Retain only one fruit for each plant.

- Harvest when the fruit is fully mature, while the stems and leaves turn brown and start to wilt.

White-skin wax gourd

- It creeps on the ground and does not need a trellis.

- No pruning is needed. Cover nodes with soil to encourage growth of adventitious roots

- The fruit's surface bears a white, protective layer of powder-like substance. When it starts fruiting, lay hay under the fruit for protection.

- Harvest when the fruit is fully mature, while the stems and leaves turn brown and start to wilt.

南瓜 Pumpkin

南瓜的品種類別繁多，本地農民經常栽種的「牛腿瓜」，是粗放式的類型。農民如有一些多出的農地，往往會在上面種植一些南瓜，這些南瓜不需經常打理，收成也未必很豐富，但即使這樣，也勝於任由土地荒棄。

不過，並非所有南瓜品種都是粗生的。如果種植的是較為精緻的品種，打理上就不能太疏懶了。

南瓜與其他瓜類的不同之處，在於其不定根系統異常發達。一般來說，當瓜蔓的節位觸及泥土，便有機會萌發不定根。不過，對大多數瓜類來說（冬瓜例外），植株的根系大體還是依靠植株的主根系的（比例佔總根系八成以上）。南瓜有趣之處，在於「瓜蔓爬到那裏，不定根就長在那裏」，不定根的比例可以超過總根系五成。這個特徵，解釋了為甚麼一棵南瓜，就可以長滿一大幅農田。

亦因為南瓜的生長太過迅速（甚至是霸道），它不太適宜在盆栽內種植。不過，外國也有一些體積小巧的南瓜品種，作園藝或觀賞用途。

There is a diverse variety of pumpkins. The tough "ox-leg pumpkin" is a common produce of local farms. Although not highly productive, it requires little care. Farmers grow it in the fields that they have no time to manage, rather than leaving the farmland abandoned. Nevertheless, some varieties of pumpkin do require more regular management input. Compared to other gourds, pumpkin has an exceptionally extensive adventitious root system. It develops wherever its lateral shoots touch soil. For most gourds, the main root is dominant and account for 80% of the root system. For pumpkin, adventitious roots account for over 50% of its root system when there is sufficient space. This explains why only one pumpkin plant can spread over a large farming field.

Due to its fast and vigorous growth pattern, pumpkin is not suitable for pot-planting. There are small-size foreign varieties of pumpkins for gardening and ornamental use.

 2 月下旬-4 月
Late February to April

 雨水
Spring Shower

深（>20）
Deep

90-120 日
90-120 Days

150-200 厘米
150-200 cm

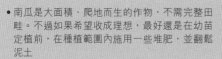

種植貼士
Growing tips

- 南瓜是大面積、爬地而生的作物，不需完整田畦。不過如果希望收成理想，最好還是在幼苗定植前，在種植範圍內施用一些堆肥，並翻鬆泥土
- 當瓜蔓跨越了種植範圍，便將其撥回田內，但要避免傷及瓜蔓及節位的不定根
- 可在果實下面墊一些乾草作保護
- 果實充份成熟，果皮堅硬時收成

- A large farming field is needed. Plough the soil and apply fertilizer before transplanting of seedlings.
- Push overgrown lateral shoots back; avoid damaging the adventitious roots.
- Lay hay under fruits for protection.
- Harvest the fruit when it is fully mature, with a hard and tough rind.

葫蘆瓜 Bottle Gourd

葫蘆瓜由於形狀趣緻，種植又不困難，常見於新界農田。除了食用，葫蘆瓜果實老身、乾燥時，瓜皮會變得堅硬。種子藏於其中，搖擺時發生「咚咚」聲響，猶如樂器。乾葫蘆可用作容器或工藝品。

除葫蘆形外，它也有較為直身的近親品種，稱為蒲瓜。

The Bottle Gourd possesses a unique shape and is easy to grow. It is commonly seen in HK farms. The Bottle Gourd's skin hardens when it gets aged and dry, and its seeds make the sound of a musical instrument when shaken. Apart from being consumed as food, dried bottle gourds are also used for making containers or in handicrafts. Clavated calabash is a kindred variety of bottle gourd but it is straight in shape.

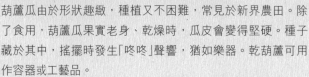 2月下旬-4月
Late February to April

雨水
Spring Shower

深(>20)
Deep

60-70日
60-70 Day

60-90厘米
60-90 cm

種植貼士 Growing tips

• 保留2-3側蔓，其餘剪除

• 由於其枝葉繁茂，可種植作為蔭棚

• 當果實仍然翠綠、光滑時，便要及時收成

• Keep only 2-3 lateral shoots and remove the rest.

• Because of its bushy nature, it is often planted as a sun shelter in summer.

• Harvest the fruit when it is fresh green and smooth.

佛手瓜 Chayote

佛手瓜的生長形態，與其他瓜類頗不相同。首先，它是少數多年生的葫蘆科作物，寒冬時植株枝葉枯萎；但翌年春天回暖時，又會重新抽發枝條，生長週期可長達數年。

另外，佛手瓜的繁殖方法也很特別，每個果實有一顆種子。將一個老身的佛生瓜埋在地下，便能長出新的植株。所以，一般售賣菜種的地方，都沒有佛手瓜的種子售賣；至於在街市購買得來的佛手瓜，往往又不夠老身。如果有相熟朋友或農場栽種佛手瓜的話，可以相託保留一些老身的果實。

除了果實，佛手瓜瓜蔓的嫩梢也可以吃，稱為「龍鬚菜」。但要注意，摘梢會刺激植株抽發新側蔓，導致養份分散；所以，如果經常採摘嫩梢，對果實的收成會帶來影響。

Chayote is an unusual crop in the Cucurbitaceae family: it is perennial and new shoots emerge in spring; rather than sowing with seed, it is propagated by planting a germinated, ripe fruit. Seeds of chayote are not available from the market. If you have a farmer-friend, ask him or her to reserve a ripe fruit for you for planting.

Beside the fruits, young shoots of chayote are also edible. Limited picking of the young shoots will stimulate the growth of lateral shoots and draw up more nutrients from the soil; frequent picking of young shoots affects the fruit harvest.

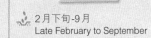
2月下旬-9月
Late February to September

雨水
Spring Shower

深（>20）
Deep

100-150日
100-150 Day

150-200厘米
150-200 cm

種植貼士 Growing tips

- 由於其枝葉繁茂，可種植作為蔭棚。只需兩株便能長滿一個10平方米的棚子。種植位置需有深厚泥層，讓其根部有足夠發展空間
- 除了11月至1月的寒冷日子，基本上大半年時間都可以栽種佛手瓜。經過冬季休眠後，翌年5月左右可再有收成
- 定時清理枯老的瓜蔓，刺激新梢成長

- Bushy, can be used as a shading trellis. Only two plants are sufficient to cover a 10m² trellis. Plant the crop on a thick soil bed near the supporting pole of the trellis to encourage root development.
- Chayote can be planted all year round except from November to January. Harvesting starts in May and spreads through the rest of the year.
- Regular pruning of old lateral shoots and wilted stem stimulates growth of new shoots and enhances yield.

翠玉瓜 Zucchini

近十數年才引進香港的瓜種，種植方法與其他瓜類大異其趣。大多數葫蘆科的農作物喜溫，故春播；但翠玉瓜則喜涼，需秋播。此外，翠玉瓜屬叢生的作物，不抽蔓，故無需搭棚。

Zucchini was introduced into Hong Kong in recent decades. Its growing requirement is different from general Cucurbitaceae crops: it likes cool weather and has to be sown in autumn; it does not develop lateral shoots and no trellis is required.

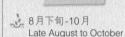
8月下旬-10月
Late August to October

處暑
End of Heat

深(>20)
Deep

60-70 日
60-70 Day

45-60 厘米
45-60 cm

種植貼士
Growing tips

- 翠玉瓜果實的甜度較高，病害也較多，種植前可先在田裏施用一些熟石灰。此外，也要定時清理田間的腐果、爛葉

- Zucchini's sweetness makes it prone to pest damage. Apply lime on soil before planting. Remove rotten fruit and wilted leaves from soil.

豆類
Peas and Beans

豆的品種五花八門，但大體而言，可分為"Bean"及"Pea"兩大類別，前者喜溫，宜春播；後者喜涼，一般在9月秋分後才開始播種。

蔬菜範疇下的豆科作物，多屬一年生植物（當然也有例外），但品種之間的栽種條件差異甚大，較大型品種如豆角，其竹棚呎吋之要求與青瓜相仿；但也有一些細小的豆科品種，能容納於4、5吋直徑的小盆中，是城市農夫的不錯選擇。種植前，需要先瞭解清楚有關品種的特性。

在永續農業上，豆科還提供另外一個重要功能，就是供應氮肥。許多豆科植物的根部，都會有根瘤菌共生。根瘤菌能將空氣內的氮，轉化成植物能吸收的氮肥，非常環保。7、8月在過渡田（即春播作物已結束，但秋播時機又未至）種植一些綠豆作為綠肥，在未開花前將其砍斷，埋入泥中，一來可提高泥土的有機質，二來也能透過綠肥的根瘤菌增加土地養份。

Beans and peas are two broad categories; the former like warm weather and should be sown in spring; the later grows in cool weather and is sown after "Autumn Equinox" in September.

Edible leguminous crops are usually annual but their growing requirements vary between varieties. Larger-sized crops like string beans require a trellis of a size similar to that of growing cucumber. Smaller-sized crops can be planted in pots with a diameter of 10-12cm and are popular in city farming.

These Leguminous crops have an important role to play in sustainable farming. The ability of the nitrogen-fixing bacteria at their root-nodule to utilize nitrogen in the soil means that they require little feeding, and also if you dig the plants into the soil after harvesting, the nitrogen becomes available for the next crop. Crops like mung bean can also be planted as green manure during fallow periods in July and August before autumn-sowing. By chopping and burying the plants in the soil before they flower, one increases the organic matter and nutrients in soil.

育苗：豆類的幼苗生長較快，發芽率亦高。一般情況下，苗棚育苗可較露地播種提早3-4星期。

施肥及護根：豆科作物的枝葉不及瓜類茂密，加上植株本身具備根瘤菌，因此豆類作物的耗肥量遠不及瓜類。田畦表面可以覆蓋護根、免掘方法來保護泥土，省水省肥。

病害：豆類作物的前期生長階段，病害不算嚴重；到了後期，會有鏽病（主要是豆角、玉豆）、白粉病（主要是荷蘭豆、蜜糖豆）等病害出現。可透過棚架或鋪乾草，避免結果時豆莢觸及濕泥面而變壞。為減少病蟲害對收成的影響，播種期不宜太遲。此外，豆科作物忌連作。

Nursery: Leguminous crops have a high germination rate and seedlings grow fast. Generally, nursery sowing can be started three to four weeks earlier than outdoor sowing.

Fertilizer application and mulching: Compared to Cucurbitaceae crops, leguminous crops grow with less stems and leaves. Aided by their root-nodule bacteria, leguminous crops require less fertilizer application. Mulching or no-dig practices can further reduce irrigation and fertilizer input.

Pest and disease: Diseases like rust (mainly in string bean and French bean) and powdery mildew (mainly in sugar pea and honey pea) are common in the later growing stage. To avoid the negative impact on harvest caused by pest damage, late sowing is not recommended with a trellis and lay hay to prevent beans and peas from touching moist soil and getting rotten. Avoid continuous cropping of Leguminous crops on the same piece of farmland.

豆角 String Bean

1. 青豆角
   String Bean (Green)
2. 白豆角
   String Bean (White)

市面常見的豆角，一般分「白豆角」和「青豆角」兩類，兩者種植時機及方法大致一樣；另外也有稱為「八月豆」的晚栽品種。

The main varieties are white string bean and green string bean. The planting method is similar. A late-sowing variety named 'August asparagus bean' is also available in the market.

 2月下旬-6月
Late February to June

 雨水
Spring Shower

 中（15-20）
Medium

 50-60日
50-60 Day

 40-50厘米
40-50 cm

種植貼士 Growing tips

- 以堆肥、骨粉為基肥，抽蔓前追施氮肥（如麩粉）
- 可分批下種以延長收成期
- 種子輕微突出，豆筴仍直身時，便要及早收成
- 需設約2米高之竹棚或籬芭

- Use compost and bone meal as base fertilizer. Add nitrogenous fertilizer before growth of lateral shoots.
- Extend harvesting period by planting different batches of seedlings at different times.
- Harvest when the pod is still straight with slightly swollen beans inside.
- Set up trellis with 2m long bamboo poles or fence.

玉豆 Snap Bean

玉豆是一個比較籠統的稱呼，不少人將短身豆筴的品種都稱為玉豆、四季豆或菜豆，外國品種的法國豆在外型上也相類似。所以，種植玉豆時，要先弄清楚相關品種的特性，因為不同品種之間的生長習性（如植株大小、生長週期、需否上棚等）可以有很大的差別。

不過，大體而言，玉豆的種植不算太困難。由於它體積較豆角小巧，故更容易融入在家居農圃之中。

Snap bean is a general name for different varieties of short-bean including French bean. Since crop size, life cycle and other growing requirements vary between varieties, it is important to identify the variety clearly before planting.

Due to its smaller plant-size than string bean, Snap Bean is more suitable for growing in a home-garden.

2 月下旬 -9 月
Late February to September

雨水
Spring Shower

中（15-20）
Medium

30-60 日
30-60 Day

15-40 厘米
15-40 cm

種植貼士 Growing tips

- 由於生長期短，種植編排可以較有彈性，9 月入冬前仍可以播種
- 部份品種需具 1-2 米高的棚架
- 種子輕微突出、豆筴仍直身時，便要及早收成

- Growing period is more flexible than string bean as its life cycle is shorter. Sow before September and avoid planting in cold months.
- Some varieties require trellis of 1-2m high.
- Harvest when the pod is still straight with slightly swollen beans inside.

荷蘭豆/蜜糖豆 Sugar pea/ Honey pea

1. 蜜糖豆
   Honey pea
2. 荷蘭豆
   Sugar pea

兩者的種植條件相類似。但荷蘭豆長粉紅花，而蜜糖豆則長白花。

荷豆/蜜豆的根系、枝葉並不發達，容易引入盆栽花園之中。可以説，一個3呎長型花盆配上一個簡單竹籬笆，已經是理想的種植環境。但要注意，荷豆/蜜豆的產量頗低，一幅2平方米的農地，種植豆角時可以有2、3公斤收成；但栽種荷豆或蜜豆時，大抵只能採得數百克。

The growing requirements for these two varieties are alike. Sugar bean's flowers are pink while honey bean's flowers are white. Their root-system is not extensive and they grow with a single shoot and are therefore, suitable for planting in pots. The yield is low compared to string bean. For a $2m^2$ piece of land, the yield of string bean can reach 2-3kg, while that of honey pea and sugar pea is only a few hundred grams.

9月下旬-10月
Late September to October

秋分
Autumn Equinox

中（15-20）
Medium

60-70日
60-70 Day

行距40-50厘米
株距15-20厘米
40-50 cm between rows
15-20 cm between plants

種植貼士 Growing tips

- 翌年3、4月遇上潮濕氣候時，荷蘭豆/蜜豆很容易染上白粉病，太遲下種會影響收成
- 以單枝幹生長，沒有側枝，可採約1米高的垂直棚架
- 種子輕微突出、豆莢仍直身時，便要及早收成

- Prone to powdery mildew damage under humid weather in March-April.
- Requires a 1m tall bamboo vertical trellis or bamboo poles for support.
- Harvest when the pod is still straight with slightly swollen peas inside.

豆苗 Pea shoot

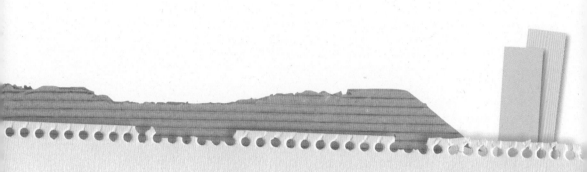

豆苗是蜜糖豆的近親，兩者的種植條件類同。但由於收成部份不一樣，因此栽種方法也有很大分別。

豆苗的栽種不算困難，但產量很低，即使種了100株，每次的收採量也不到500克；而採摘豆苗的工夫也頗為費時，種植者應先作考慮。

Pea shoot is a kindred variety of honey pea. They share similar growing requirements but are different in growing methods, as different parts of the plant are harvested.

Pea shoot is easy to grow, but its yield is fairly low - only 500g of pea shoot can be harvested, each time, from 100 plants. Besides, harvesting pea shoots is time consuming. These factors should be considered before planting.

	9月下旬-10月 Late September to October
	秋分後 Autumn Equinox
	中（15-20） Medium
	40-50日 40-50 Day
	行距30厘米 株距10-15厘米 30 cm between rows 10-15 cm between plants

種植貼士 Growing tips

- 由於收採的是嫩梢部份，因此種植編排、施肥應採葉菜類的形式來管理。豆苗是連續收採的作物，生長中段需要補施氮肥。種植環境理想的話，同一幅豆苗田可收採5至6次

- 豆苗是單枝生長的植物，待枝條長至20-25公分長時，便可將頂端的嫩梢收採。收採後，折斷位下方的葉筴會長出新枝條，隔三星期左右便能再次收採

- 避免重複在同一個節位採摘豆苗，因為同一組葉筴難以承受萌生太多的新梢

- Because young shoots are harvested, management, including planting arrangement and application of fertilizers, should be similar to that of growing leafy vegetables. As pea shoot can be harvested continuously, nitrogen fertilizer should be replenished in the mid-growing stage. 5-6 times of harvest are available if the growing condition is ideal.

- Pea shoot grows with a single shoot. Harvest the top part when shoot reaches 20-25cm. A new shoot grows right below the reaping point and another harvest can be made three weeks later.

- Avoid harvesting at the same point as it inhibits growth of a further new shoot.

茄科
Solanaceae Crops

茄科的農作物主要包括番茄、茄子和椒，三者各具多個不同形態的品種。由於三種農作物的外貌並不相像，不少人不知它們原來是關聯的農作物。

茄科作物多忌積水，喜全日照，光線不足會導致少花少果。連作是茄科的大忌，種植過茄科作物的泥土，兩年內不宜再種植同類或類近的品種。

Common Solanaceae crops include tomato, eggplant and pepper. Although their fruits vary in shape and size, they all need plenty of sunlight and soil with good drainage.

Avoid continuous cropping of Solanaceae crops, of either the same or different varieties, within two successive years.

番茄 Tomato

8月下旬-10月	Late August to October
處暑	End of Heat
深(>20)	Deep
40-90日	40-90 Day
30-60厘米	30-60 cm

番茄乃一年生作物，主要分肉茄及車厘茄兩大類。顏色方面，番茄以紅色為主，但也有黃色、朱古力色的品種，種植條件大同小異。

番茄有分為「自封頂」及「非自封頂」兩種形態，後者可不斷抽發新的枝條，因此可採用單枝幹型的栽種形式，具體方法是以直身棚來栽種，剪除所有側枝，讓植株垂直而生。至於自封頂型的品種，生長週期較短，收採期亦會較集中；不過，可透過適量修剪來延長收成期。

Tomato is an annual crop, which can be divided into pulp tomato and cherry tomato. Red tomato is common and it is also available in yellow and chocolate colours. All varieties share growing requirements. Tomatoes are obtainable as bush or upright (climbing) types. Plant the upright type with a vertical trellis, remove all of its lateral shoots often, and let the main stem climb upward. The growing and harvesting period of the bush varieties are shorter. Proper pruning extends the harvesting period.

種植貼士 Growing tips

- 定植後在田面鋪上護根，保護泥土及幼苗。茄科的作物也適合用於免掘種植的方法

- 番茄喜涼，不耐濕熱。無論是何時播種，到了翌年4、5月時，番茄的長勢都會呈現下滑趨勢，最終死亡。所以，番茄秋播的時機宜早不宜遲。不過，8月由於天氣仍然太熱，可先在半蔭棚內育苗，苗期約一個月。

- 堆肥、骨粉等可在基肥時使用，幼苗定植後數天可追施氮肥。進入收成期時需減少施用氮肥

- 較細小品種可直接以短竹扶持主幹；至於較大的品種，則應以竹棚環繞植株外圍，再以繩子套上主要枝幹

- 棚底種植者，可選種非自封頂型的品種，透過剪除側枝，讓植物垂直而生

- 修剪乃重要一環。肉茄由於果實大，需較粗壯的結果枝來支撐，所以植株只保留最壯健的一、兩條側枝；車厘茄的果實小，相對來説可以保留較多側枝，但也不宜過多

- 果實呈橙紅色時便可以收成

- Apply mulch after transplantation to protect seedlings and soil. No-dig practice is suitable for Solanaceae crops.

- Tomato cannot tolerate humid and hot weather. Nurse seedlings in a semi-shaded shelter in August and it takes one month for the seedling to get ready for transplanting. The crop wilts in May of the next year, regardless of the sowing time.

- Reduce use of nitrogen fertilizer during the harvesting period.

- For small-size varieties, support the main stem with short bamboo poles. Trellis is needed for large-size varieties, tie the main stem loosely to the trellis.

- Choose upright varieties for growing under shelter. Remove lateral shoots to keep the plant growing upright.

- Requires proper pruning. For pulp tomato varieties, keep only one or two strong lateral shoots. For small-size cherry tomatoes, more, but not all, lateral shoots can be kept.

- Harvest when the fruit turns orange red.

茄子 Eggplant

春播：2月下旬 - 4月
秋播：8月下旬 - 9月
Spring: Late February to April
Autumn: Late August to September

春播：雨水
秋播：處暑
Spring: Spring Shower
Autumn: End of heat

深(>20)
Deep

80-90日
80-90 Day

50-60厘米
50-60 cm

茄子的植株結構與番茄類近，但喜溫，生長週期較長，可作兩年生。

現時市面的茄子以紫色、白色為主，但也有青色、帶花紋的品種。一般而言，白色的茄子較容易出現病害。

Eggplant has a similar plant structure to that of the tomato. It grows in warm weather, has a longer life cycle, and can grow for two years.
Eggplant is available in purple, white and green. White eggplant is more prone to disease.

種植貼士 Growing tips

- 泥土與肥料的管理方法，與番茄相類同
- 春播的茄子約7月收成；秋播則在翌年3、4月收成
- 茄子不宜露地播種，宜以帶穴苗盤育苗。苗期長，需約要兩個月時間
- 茄子的植株結構較番茄堅固，但最好也以竹枝扶持主幹，避免大雨時塌下
- 只保留最壯健的一、兩條側枝。如果作兩年生，生長中段需勤加修剪，並除去弱枝、短截禿枝，避免果實變小，甚至不結果
- 果實仍然光滑、有彈性時，便要及早收成

- Soil preparation and fertilizer application are similar to that of tomato.
- The Spring-sowed plant is ready for harvest in July while autumn-sowed one is ready in the following March and April.
- Use seed tray with hole or seedling pot for propagation. Seedlings need two months to be ready for transplant.
- The structure of eggplant is stronger than that of tomato. Use bamboo poles to support the plant.
- Keep one or two of the strongest lateral shoots. For planting as perennial, remove weak stem and short bare stem regularly.
- Harvest when the fruit is still smooth and springy.

椒 Pepper

1. 甜椒
Sweet Pepper

2. 辣椒
Hot Pepper

甜椒（Bell Pepper）

2月下旬-4月
Late February to April

雨水
Spring Shower

深（>20）
Deep

80-90 日
80-90 Day

40-50 厘米
40-50 cm

辣椒（Hot Pepper）

春播：2月下旬-4月
秋播：8月下旬-9月
Spring: Late February to April
Autumn: Late August to September

春播：雨水
秋播：處暑
Spring: Spring Shower
Autumn: End of heat

深（>20）
Deep

100-120 日
100-120 Day

30-40 厘米
30-40 cm

分辣椒（右圖）及甜椒（左圖）兩大類。辣椒一般果實較小，生長週期較長，可存活2、3年；甜椒的果實較大，生長期較短，僅一年生，忌積水，枝幹較弱，總體來說較辣椒難種。所以，部份人會選擇在透光棚底下種植甜椒，以減低暴雨的影響，這樣存活率明顯較高。

無論是形狀及顏色，椒的品種都五花八門。顏色方面有紅、燈、黃、綠等，形狀則有長條形、肥短形、燈籠形、圓形等，也有一些以觀賞為主的園藝品種。

Pepper is divided into two main types. Hot pepper and bell pepper. Hot pepper is generally smaller in size and has a longer life cycle, up to 2 to 3 years. Bell pepper is an annual crop with a shorter life cycle. Its stem is weaker and it cannot stand a water-logged environment. Bell pepper is more difficult to grow than hot pepper. Farmers usually grow bell pepper under a transparent shelter for protection against rainstorms.

Pepper is diverse in shape and color. It is available in red, orange, yellow and green. As for shape, it is available in slender, dumpy, lantern-shaped and round. Some peppers are grown for ornamental purposes.

種植貼士 Growing tips

辣椒

- 多年生，可春播或秋播；無需棚架
- 到了收成期中段，則要透過修剪、短截、去除老枝等方法，延緩植株老化
- 果實呈橙紅色時，便可收成

甜椒

- 一年生，只能春播
- 植株較高，結構又不及茄子壯健，故容易塌下，需以枝架來扶持
- 一般無需修剪
- 果實仍然光滑、有彈性時，便要及早收成

Hot Pepper

- Perennial and can be sown either in spring or autumn. No trellis is needed.
- Pruning, shortening and removal of old branches are needed.
- Harvest when the fruit turns orange red.

Bell Pepper

- Annual and should be sown in spring.
- A trellis is needed to support the taller and weaker plant structure.
- No pruning is needed.
- Harvest when the fruit is still smooth and springy.

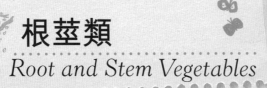

根莖類
Root and Stem Vegetables

根莖類泛指一些以根部或莖部作為食物部份的農作物。由於這個分類並沒有很清晰的標準，所以各種作物之間的生長習性，也沒有明顯共通性。

許多人會混淆蘿蔔和甘筍這兩個品種。其實，甘筍屬傘形科，與屬於十字花科的蘿蔔全不相干，兩者只是收成部份的模樣有些相似。

This group of vegetables are crops of which either the stem or root are consumed. The harvest of carrot and radish look alike but they actually belong to Apiaceae and Brassicaceae families respectively. The growing methods and characteristics vary among different varieties.

蘿蔔 Chinese Radish

早水：8月下旬-9月
遲水：10月-11月
Early sowing: Later August - September
Late sowing: October to November

早水：處暑
遲水：寒露
Early sowing: End of heat
Late sowing: Cold Dew

深(>20)，櫻桃蘿蔔可用淺畦
Deep (shallow for cherry belle radish)

70-110日
70-110 Day

20-35厘米
20-35 cm

品種包括白蘿蔔、青蘿蔔，還有體型小巧的櫻桃蘿蔔。當中，以白蘿蔔的栽種最為普遍，並可再細分為早水、遲水等品種，後者的體型較大，生長期亦較長。

White Chinese radish, green Chinese radish and the small-sized cherry belle radish are common varieties. White Chinese radish is easy to grow and popular. It is sub-divided into early-sowing and late-sowing varieties, for which the latter has a longer growth period and the produce is larger in size.

種植貼士 Growing tips

- 蘿蔔進入根部膨脹時期，一方面會向下深入泥層，另一方面也會向上生長。所以，蘿蔔的田畦宜採用「打巷」形式(見圖)，方法是在播種(行播)的範圍挖一條5-10厘米的淺坑(深度視乎品種而定)，泥土墩在兩邊，種子種在坑內。當蘿蔔膨大、突出田面時，逐步將兩邊的泥土回填坑中，掩蓋蘿蔔伸出的部份

- 不能採培苗方法，必需露地播種，因為移植容易令蘿蔔植株的主根受傷

- 如果是製作蘿蔔糕，應採遲水種，並以翌年農曆新年計算，倒數100-110天作為播種日子

- 播種後數天即發芽，生長初段可勤澆水以保持泥土濕潤。當蘿蔔開始膨大時，必需適量制水，因為過濕環境容易引發病害，或令蘿蔔裂開

- 開花之前收成

- Both ends of white Chinese Radish swell and elongate when it grows. When the top part emerges on the soil surface, cover it with soil. See figures below for soil preparation techniques.

- Outdoor sowing is preferred to avoid root damage.

- For making radish cake, choose late-sowing varieties and sow at 100-110 days before Chinese New Year.

- Seed germinates within a few days after sowing. Frequent irrigation is needed to keep soil moist during early growth stage. When the root starts to swell, regulate the irrigation. Water the crop when soil is dry. Otherwise, root splits because of excessive water content.

- Harvest before flowering.

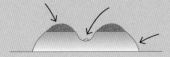

甘筍 Carrot

🌱	9月-2月 September to February
🔍	白露 White Dew
🐸	中（15-20） Medium
🌾	60-70日 60-70 Day
⚘	5-20厘米 5-20 cm

中國人統稱的「紅蘿蔔」，其實並非蘿蔔（十字花科）一類，甘筍才是其真實身份。較幼身、爽脆的甘筍品種，多作沙律等鮮食，較粗大的品種則常用於其他煮食用途，如湯類。

種植甘筍不算困難，病害也不多，但部份工序會比較花時間。甘筍必需露地播種，每株保持5-20厘米的生長空間（視乎品種而定），種植量多時可採撒種方式，並注意疏苗、除草等工作。保持田面水平，避免種子堆積一處，每平方呎的種子用量約為100粒。

The produce which is commonly named as "Red Radish" by Chinese is not categorized as radish. It is actually classified as "carrot". The slender and crispy varieties are often consumed in salad, while the larger varieties are often used in soup.

Carrot is easy to grow and not prone to pest and disease, but it requires more management input in thinning and weeding. The distance between plants should be kept around 5-20 cm(depending on species). It has to be sown outdoor and broadcast sowing can be used for planting in large quantity. Keep the surface flat and even, and avoid any accumulation of seeds. A square feet of soil should contain about 100 seeds evenly scattered.

種植貼士 Growing tips

- 可分批種植，以延長收成期
- 不能採育苗方法，因移植容易令主根受損，令甘筍開叉。整理田畦後，可直接在田面撒播種子；如擔心撒播不平均，可先用幼砂粒與種子混合。對於種植量較少的園藝種植者，則可考慮採用行播方法
- 播種後一個月內，需勤澆水、除雜草以及疏苗，保持植株前期階段能順利成長。植株穩定後，後期的管理工作便會簡單得多
- 與白蘿蔔不同，甘筍根系主要向下伸延，故田畦無需打巷；即使到收成階段，根部也只會有小部份露出田面
- 拔取收成時要小心避免折斷

- Different batches over time can extend the harvesting period.
- Direct outdoor sowing is preferred to avoid root damage. Mix seeds with fine sand for sowing more evenly. For growing a small quantity at home, sowing in line is recommended.
- Frequent irrigation, weeding and thinning are needed within a month after sowing. Less input is needed afterwards.
- Different from radish, carrot's root extends downward. The top part of carrot exposes on the surface when it is ready for harvest.
- Harvest by pulling it out carefully,

蕪菁 Turnip

不少人會混淆蕪菁及蘿蔔，其實兩者雖同屬十字花科，但算不上近親；蕪菁的收成部份較短，顏色較多樣化(有紅、粉、紫等顏色)，根部膨脹時主要向上伸延，不似白蘿蔔般同時向上下伸展。

Turnip is different from Chinese radish although both of them belong to the Brassicaceae family. Compared to the Chinese radish, turnip is shorter, available in various colors (eg. red, pink and purple), and its root swells and grows upward.

9月-12月
September to December

白露
White Dew

中(15-20)
Medium

70-90日
70-90 Day

20-30厘米
20-30 cm

種植貼士 Growing tips

- 種植前先清理雜草
- 可分批種植以延長收成期
- 可育苗或直接露地播種，但無需打巷
- 管理方法與白蘿蔔類近
- 於開花前收採

- Remove weeds before planting.
- Plant the crop in different batches over time to extend the harvesting period.
- Propagate either at nursery or direct outdoor sowing.
- Management work is similar to that of Chinese radish.
- Harvest before flowering.

紅菜頭 Beetroot

常見於西式菜譜，早年本地的農民較少栽種。不過，近年它已成為相當流行的根莖類作物。除了紅色，也有黃色的金菜頭，種植方法大致類同。其實，紅菜實的葉片也可以採食，味道、質感與君達菜相類似。

Beet Root is a common ingredient in western cuisine and has become a popular crop of local farms only in recent years. The growing requirement of red and yellow beet root is similar. Leaves of beet root are also edible and taste like Swiss chard.

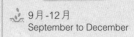

9月-12月
September to December

白露
White Dew

中（15-20）
Medium

120 日
120 Day

15-20厘米
15-20 cm

種植貼士 Growing tips

- 種植前先清理雜草
- 根系較弱，不利移植，可以帶穴苗盤育苗，或採露地行播
- 管理方法與甘筍類近
- 前期工夫較多，生長穩定後便無需頻密管理
- 可分批種植以延長收成期，最遲12月播種
- 鮮吃的紅菜頭宜早收採（上部葉片尚未枯萎時），煮湯用的則可遲一些收成

- Weed before planting.
- Root system is weak. Use seed tray with cells for nursery, or sow directly outdoor.
- Management work is similar to that of carrot.
- More input is needed in the early stage.
- Sow different batches over time to extend the harvesting period. Sow by December.
- For fresh consumption, beet root should be harvested early (before upper leaves wilting). For soup serving, later harvest is allowed.

芥蘭頭 Kohlrabi

顧名思義，與芥蘭是親屬。但由於收採的是莖部，故管理方法有所不同。現時，本地栽種的芥蘭頭主要分青、紫兩隻顏色，種植方法基本上一樣。

芥蘭頭是相對容易栽種的農作物，管理也不麻煩，是秋播的理想選擇。

It is a kindred variety of Chinese kale but unlike kale, its stem is consumed. Green and purple varieties share similar growing requirements and they are commonly grown in winter. It is rather easy to grow.

8月下旬-12月
Late August to December

處暑
End of Heat

深（>20）
Deep

70-80日
70-80 Day

30-40厘米
30-40 cm

種植貼士 Growing tips

- 可分批種植以延長收成期，但不要遲於12月
- 與蘿蔔等根莖類作物不同，芥蘭頭不存在「主根開叉」問題（芥蘭頭收成的並非根部），故無需露地播種，採育苗方法的效率會較高
- 定植後，苗的四周可以護根覆蓋，護苗保水
- 枝葉枯萎前收成

- Sow different batches over time to extend the harvesting period. Sow by December.
- Nursery enhances growing efficiency.
- Add mulch after transplanting for seedling protection and keeping soil moisture.
- Harvest before wilting of stems and leaves.

雜糧
Field Crops

雜糧泛指一些種植週期較長、所需空間較大、非主流蔬菜的品種類別,例如番薯、芋頭、薑等。由於雜糧的市場價值較低,也無需頻密管理,所以農夫多選擇在一些質素較次、或位置不太方便的地方來種植。

Field crops refer to agricultural crops grown over a large area. Local field crops such as sweet potato, taro and ginger usually have a long growing period and require little management input. Due to the low market price, farmers often plant field crops on farmland of lower soil quality or reduced accessibility.

育苗:許多雜糧都採用營養繁殖的方法,較少使用種子。由於繁殖體本身已累積一定養份(如薑、芋等),發芽一般會較順利,未必需要採用育苗的手段。

施肥及護根:雜糧的管理工作較少,不建議頻密施肥或灌溉。定植前可先施入充足基肥,定植後採大量護根覆蓋田面,這是保水、保肥、防止雜草生長的良好方法。

Nursery: Asexual propagation is common in field crops. A nursery is not required and the crops like ginger and taro often propagate and spread easily.

Fertilizer application and mulching: Field crops generally require little management input in terms of fertilizer application and irrigation. Apply sufficient base fertilizer before cropping. Adding mulch after burying the vegetative parts will help to retain water, preserve fertility and control weeds.

番薯 Sweet Potato

香港位於亞熱帶，種植薯仔的效益相對低一些，種植番薯則非常適合。

傳統的番薯以紫心、黃心兩類為主，但近年也引入了一些體積較小、或專門採葉的品種。番薯是不適宜新鮮即吃的，收成後放在蔭涼地方一、兩星期，有助糖份累積，食用價值會較高。除了收採根部，番薯本身也是華南地區廣泛採用的覆蓋植物。不少農民喜歡在荒地上種植番薯，植株的蔓、葉可以用來餵飼豬隻。

As Hong Kong is located in a sub-tropical region, it is more suitable for growing sweet potato than potato.

Sweet potato of purple and yellow pulp used to be the most common varieties in the local market. New varieties of smaller size or varieties that are just for leaf production were introduced in recent years. Be sure of the variety that you need before planting. Rather than consuming fresh after harvest, storing sweet potatoes in a shady environment for up to two weeks enhances their sweetness. Sweet potato is also widely planted as a cover crop in Southern China. Farmers grow it on barren land and harvest shoots and leaves as pig feed.

🌱	春：3月 - 4月 秋：8月 - 9月 Spring: March to April Autumn: August to September
🔍	春：驚蟄 秋：處暑 Spring: Insects Waken Autumn: End of Heat
🌿	深(>40) Deep (>40)
🌼	180-240 日 180-240 Day
🌱	30-40 厘米 30-40 cm

種植貼士 Growing tips

- 宜採高畦、利疏水的圓曲頂型「龜背田」。種植前可先施用堆肥、骨粉等作為基肥
- 番薯一般以插枝繁殖。由於它廣泛流行於各鄉村及社區農圃，要取得枝條應該並不困難
- 插枝的枝條最少有五、六個節位，其中下面三、四個節位需剪去所有葉片，斜插入泥，讓其發根(見圖)；上方、露出田面的兩、三個節位可保留小量葉片。節位未發根前需澆水以保持泥土濕潤，生長穩定後便無需灌溉，太濕反而容易出現病蟲及鼠害。番薯無需經常打理，只需枝蔓抽生至種植範圍外時，才將其撥回田內
- 春植者三、四月種植，入冬前收成；秋植者八、九月種植，入冬前成形，翌年五月收成
- 收成時，在田畦側邊翻土，小心避免掘傷番薯
- 由於根部藏於泥土內，表面看不見，需根據種植的日數來決定收成時間，春植約需200天，秋植則240天

- Plant on high ridges of well-drained soil. Apply compost and bone meal as base fertilizer before planting.
- Sweet potato is usually propagated by cutting. As the crop is common in villages and community farms nowadays, it is not difficult to get the cuttings.
- Prepare cuttings with strong stems that bear at least five nodes. Clear the leaves from the four lower nodes. Plant the cuttings in the soil aslant in a single row and expose at least two upper nodes on the soil surface. Water frequently to keep soil moist until the root system is established. Reduce irrigation when the crop stabilizes to reduce pest damage related to humidity. Put lateral shoots back if they grow beyond the ridge.
- Plant in spring for harvesting before winter; plant in August-September for harvesting in the following May.
- Loosen soil along the ridge before harvesting to prevent damaging the sweet potatoes.
- You cannot tell when to harvest by judging its appearance as sweet potato is buried in soil. Thus, harvest by counting the growing days (200 days for spring sowing, 240 days for autumn-sowing).

木薯 Cassava

在香港，木薯的種植不算普遍。但在世界上，它是一種非常主要的農作物。它容易種植，病蟲害少，能適應較貧瘠的泥土，並可提供很高的熱量；將木薯打碎、乾燥為木薯粉後，又能製作成多種食物。近年，更有不少地方以木薯作為生物燃料。

Although cassava is not commonly grown in Hong Kong, it is a major crop globally. It is easy to grow, not prone to pest and disease, highly adapted in infertile soil, and is rich in energy content. It can be processed into tapioca which is an essential ingredient for various food products. In recent years, cassava has also been planted for supplying biofuel in some countries.

🌱	2月下旬-3月 Late February to March
🔍	雨水 Spring Shower
🌿	深(>30) Deep (>30)
🌾	240-280 日 240-280 Day
⬇	50-60厘米 50-60 cm

種植貼士
Growing tips

- 高畦，插枝處留淺坑，以利貯水

- 以堆肥、骨粉作基肥，建議採護根覆蓋田面，以保水保肥、控制雜草

- 插枝的枝條不宜老也不宜嫩，最好是直徑不少於1厘米、尚未木質化的部份。枝條最少有六、七個節位，無需留葉，一半斜插入泥(上下方向千萬不能倒轉)，一半露出田面。節位未發根時，需澆水以保持泥土濕潤；植株生長穩定後便無需經常打理

- 木薯直立而生，容易為颱風吹倒。為了減低傷害，颱風到達前，可先將較高部份的枝幹砍斷

- 收成期為11月下旬至翌年1月，需保留種源，為翌年提供插枝的枝條

- Plant cuttings in a shallow pit to enhance water holding.

- Apply compost and bone meal as base fertilizer before planting/cropping. Add mulch to preserve soil fertility and control weeds.

- Prepare cuttings that bear six to seven nodes. Cuttings should be over 1cm in diameter. Aged hardy stems or young shoots should be avoided. Clear all the leaves and plant half of the cutting in the soil aslant with the upper part pointing upward. Plant in a single row. Water frequently to keep soil moist until the root system is established. Management input can be reduced when cuttings are stabilized.

- Pruning the upper stems before the typhoon season helps reduce loss.

- The root is ready for harvested between late November and January. Preserve some mother plants for cutting supply for the following year.

芋頭 Taro

芋頭喜濕易種，適合種在較潮濕的泥土環境中。在街市所買的「芋仔」，只要外表完好、尚未枯萎，也能用來栽種。

Taro likes moist soil and is easy to grow. Any small taro available in good condition from market can be used for growing.

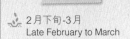

2月下旬-3月
Late February to March

雨水
Spring Shower

深(>30)
Deep (>30)

210-240日
210-240 Day

50-60厘米
50-60 cm

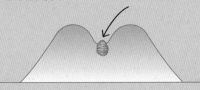

種植貼士 Growing tips

- 芋頭的體積大，且喜濕，因此田畦的設計宜採「火山」形高畦(見圖)。這地型設計能有利儲水，即使不經常澆水，也能保持坑內濕潤

- 以堆肥、骨粉為基肥。由於潮濕的環境易引來病害，施基肥時可放一些熟石灰。種植後，應以護根覆蓋田畦

- 收成期為10月下旬至12月，保留小量「芋仔」作翌年種植使用

- A ridge with a deep pit is needed for planting taro. Plant taro at the centre of a little 'volcano' with its tip exposed (see Figure below). The shape retains water and keeps soil moist without frequent irrigation.

- Apply compost and bone meal as base fertilizer before cropping. As humid soil attracts pests, apply lime as base and mulch after planting.

- Harvest between late October and December. Preserve provenance for planting.

薑 Ginger

薑是華南地區普遍栽種的農作物，用途廣泛，種植方面亦不困難。至於沙薑、黃薑等品種，種植方法與一般的薑大同小異，只是由於用途較狹，所以並不常見。除了食用薑以外，也有觀賞品種的薑花，氣味清香，適合在沼澤地方生長。

薑一般在春天種植、夏天生長、入冬時收成。7月薑身仍嫩的時候，薑可以用來直接食用，這階段稱為「子薑」；11月後枝葉枯黃、養份已充份累積在根部時，稱為「老薑」。

Ginger is easy to grow and it is widely consumed in Southern China. Varieties like sand ginger and turmeric share similar growing requirements but their use is comparatively limited. There are also ornamental varieties with pleasant fragrance that are planted in wetlands.

Ginger is generally planted in spring, growing vigorously in summer and ready for harvest in winter. In July, 'young ginger' is harvested for consuming in raw form. When the stems and leaves wilt in November, nutrient is accumulated in the root and we call the produce 'old ginger', which has a strong favour but is not suitable for consuming in raw form.

1. 薑成熟時，上部的莖葉會呈枯黃。
Stems and leaves will turn yellow when ginger become mature.

2月下旬-3月
Late February to March

雨水
Spring Shower

中（15-20）
Medium

子薑：150-180
老薑：240-270
Young Ginger: 150-180
Old Ginger: 240-270

行距30-40厘米
株距20-30厘米
30-40cm between rows
20-30cm between plants

種植貼士 Growing tips

- 將老薑截斷成一段段，每段最少保留三、四個芽位。然後將截斷的薑塊埋入泥土，儘量將較多的芽位指向上方，芽頂端尖位稍露出來（見圖）

- 與蘿蔔一樣採「打巷」式田畦，方法是在種植薑頭的位置挖一條8-10厘米深的淺坑，泥土墩在兩邊；每條坑種植兩至三株；待薑長至一定高度時，將兩邊的泥土回填坑中，覆蓋薑身。這個方法可讓薑的根部有較大發展的空間

- 以堆肥、骨粉作基肥，種植後以護根覆蓋田面

- 收成期為11月下旬至1月

- Cut ginger into several pieces, each bearing 3-4 nodes, for planting. Nodes should be facing upward with tips exposed on the soil surface.

- Use a ridge shape that is similar to that used for growing white Chinese radish. Plant 2-3 pieces of ginger at a shallow pit of 8-10cm deep. During the growth of crop, gradually refill the pit with soil from both sides of the pit to stimulate root development.

- Apply compost and bone meal as base fertilizer before cropping. Mulch after planting.

- Harvest between November and January.

沙葛 Yam bean

沙葛其實是豆科的作物，也確實會有開花結豆的階段。不過，由於沙葛的收成部份是根部，所以一般歸入雜糧類別。

Yam bean belongs to the Fabaceae family and bears flowers and pods. Its root is harvested for consumption.

2月下旬-6月
Late February to June

雨水
Spring Shower

中（15-20）
Medium (15-20)

100-150日
100-150 Day

20-30厘米
20-23 cm

種植貼士
Growing tips

- 需設簡單竹棚為沙葛的枝蔓提供空間
- 沙葛的種子含有毒性，千萬不能收採豆莢來食用。可將豆子摘除，為植株節省養份
- 以堆肥、骨粉作基肥，種植後以護根覆蓋田面
- 種植4-5個月，讓養份充份累積於植株根部
- 當枝葉開始枯黃時，便可以收成

- It bears lateral shoots and a simple trellis is needed for planting.
- Its pod is not edible and its bean is poisonous. Pick the pod when it is developed to preserve the nutrient.
- Apply compost and bone meal as base fertilizer before cropping. Mulch after planting.
- Ready for harvest in 4-5 months when nutrients have accumulated in the root.
- Harvest the produce when stems and leaves start to wilt.

西蘭花 Broccoli

過往，西蘭花曾經是「高價菜」。但由於廣泛種植、品種改良等原因，現時已變得大眾化。

西蘭花主要分為早水及遲水，前者生長期較短，花球亦較小，後者則相反。西蘭花全株只有花蕾及部份莖部屬食用部份，所以產量相對會較低。

The once pricey broccoli is now getting less expensive because of wider cultivation and crop improvement.

Early-sowing and late-sowing varieties are commonly grown by local farmers (details can be found on seed package). The former has a shorter growing period but smaller heads, while the latter is the opposite. The yield of broccoli is relatively low as it has a long growing period but only its florets and stalks are edible.

8月下旬-12月
Late August to December

處暑
End of Heat

 深（>20）
Deep (>20)

 70-100日
70-100 days

 40-50厘米
40-50 cm

種植貼士
Growing tips

- 宜採育苗方法。可分階段播種以延長收成期，但播種期應不遲於12月。

- 病害不多，但雀鳥常侵食葉片，需小心防範

- 收成日期需小心掌握，太早收採花球尚未完全發育，太遲花蕾散開則導致品質下降。收採後主薹旁會長出側芽，可延後收採一些小花蕾

- 適合以免掘方法來種植

- 先以堆肥、骨粉等混入田畦中作基肥；枝葉生長時，再追施氮肥（如麩粉）

- 當花球仍然結實、尚未散開時收成

- Nursery is preferred. Extend harvesting period by planting several batches over time. Sow by December.

- Not prone to disease. Protection from birds is essential.

- Harvest on time to avoid pre-mature florets or flowering. Broccoli continues to produce side-shoots over a considerable period, they should be removed regularly.

- No-dig method is recommended.

- Apply compost and bone meal as base fertilizer. Apply nitrogenous fertilizer (e.g. peanut cake) to support leaf growth.

- Harvest when the flower bud is still packed and not scattering.

椰菜花 Cauliflower

椰菜花是西蘭花的近親，兩者種植方法大同小異。過往，椰菜花主要以早水、遲水來區分。不過，近年引進的椰菜花品種五花八門，有橙色、紫色、白色等，又或是花蕾呈鑽石型的品種。

Cauliflower is a variety of broccoli and they share similar growing requirements. It is broadly divided into early-sowing and late-sowing varieties. Other varieties of various colors (eg. orange, purple and white) and shapes (eg. spiraled, pinnacle-shaped florets) were introduced in recent years.

8月下旬-12月
Late August to December

處暑
End of Heat

深(>20)
Deep

80-120日
80-120 days

50-60厘米
50-60 cm

種植貼士
Growing tips

- 種植條件與西蘭花類似，但需預留較多空間。另外，椰菜花不會長出側芽
- 當花球仍然結實、尚未散開時收成

- Similar growing requirement as broccoli except it requires more space between plants and it bears no side-shoots (i.e. a single harvest).
- Harvest when the flower bud is still packed and not scattering.

粟米 Corn

粟米品種的類別繁多，一些可作為主糧，也有一些是用作飼料或鮮食。甜度較高的品種一般產量較低，對水肥要求較嚴格，病蟲害也會多一些。雖則如此，由於超甜粟味道較佳，農夫或園藝栽種者多數還是選擇種植這類品種。

許多人不知道，甜粟其實是可以「生食」的。不過，粟米的糖份會隨貯藏時間延長而逐漸下降，因此，收採後需儘快食用。

Corn is diverse in variety and is consumed as a basic dietary item or animal feed. Generally the sweet varieties are low in productivity, more prone to pest, and require higher water input and high soil fertility. Nevertheless, farmers and gardeners prefer sweet varieties because of their better taste. Sweet corn should be consumed shortly after harvest or its sweetness degrades over time. It can be eaten raw.

2月下旬-4月
8月下旬-9月
Late February to April
late August to September

春：雨水
秋：處暑
Spring: Spring Shower
Autumn: End of Heat

中（15-20）
Medium

70-80日
70-80 days

30-40厘米
30-40 cm

種植貼士 Growing tips

- 宜以「打巷」形式栽種，先挖一條5-8厘米淺坑，泥土墩在兩邊，種子/菜苗種在坑內。當株植長大、突出田面時，逐步將兩邊的泥土回填坑中，掩蓋植株的根部

- 以堆肥、骨粉等混入田畦中作基肥。枝葉生長時，再追施氮肥（如麩粉）

- 育苗或露地播種均可，當中以育苗的存活率較高。粟米的幼苗生長迅速，因此定植需及時

- 每棵最多保留1-2個果實，其他幼果可及早摘下作為「粟米芯」

- 粟米容易被颱風吹倒，這時不應該強行將植株扶直，因為這樣會令植株受到「二次傷害」，可順着伏倒方向以竹枝等材料承托

- Draw a 5-8cm deep pit and put soil to both sides of the pit. Plant seeds or seedlings in the pit. Fill the pit with soil gradually as the plants grow.

- Apply compost and bone meal as base fertilizer. Apply nitrogenous fertilizer(e.g., peanut cake) to support leaf growth.

- Sowing outdoors is possible. Nursery planting will enhance the survival rate. Corn seedlings grow rapidly, thus transplant before they get too old.

- Keep only 1-2 corns per plant. Young and premature corn can be harvested.

- Avoid growing in the typhoon season. If a plant collapses after typhoon, support it with poles. Never put the damaged plant upright forcefully as this may result in further damage.

秋葵 Okra

秋葵喜溫，適合在春、夏種植。由於中式菜譜較少機會用到秋葵，因此過往種植者不多。但它其實非常合適華南地區栽種，加上夏季的作物種類較冬天少，秋葵的出現更讓炎夏的田園多了一個選擇。

品種方面，較主流的是青色果實品種，其次是紅色。兩者的種植方法差別不大。

Okra is not a common ingredient in Chinese cuisine and therefore it is not widely grown in local farms. However, warm-loving okra is very suitable for growing as a summer crop in Southern China. The growing requirements of the common green and red-skinned okra are similar.

2月下旬-4月 Late February to April	
雨水 Spring Shower	
深(>20) Deep (>20)	
70-80日 70-80 days	
40-50厘米 40-50 cm	

種植貼士 Growing tips

- 以堆肥、骨粉等混入田畦中作基肥，枝葉生長時，再追施氮肥（如麩粉）
- 苗棚育苗可讓播種期提早2、3星期，成活率亦較佳
- 8月時較多病蟲害問題出現，因此太遲種植會影響到收成期
- 可採用免掘的方法種植
- 需嫩收，收成果實保持在10厘米左右，稍有延遲果實便會硬化老化。收成高峰期時，基本上每天都要採摘
- 如被颱風吹倒，不應該強行將植株扶直，因為這樣會令植株受到「二次傷害」，可順着伏倒方向以竹枝等材料承托

- Apply compost and bone meal as base fertilizer. Apply nitrogenous fertilizer(e.g. peanut cake) to support leaf growth.
- Sowing outdoors is possible. Nursery planting will enhance the survival rate.
- Severe pest damage is likely in August. Delay in sowing shortens the harvesting period.
- No-dig method is recommended.
- Okra is ready for harvest when it reaches 10cm long. Harvest daily during the peak production period. Fruits will harden and age if there is a delay in harvest.
- If a plant collapses after a typhoon, support it with poles. Never put it upright forcefully, as this can result in further damage.

枸杞 Matrimony vine

枸杞乃多年生作物，炎熱時枯倒，氣候轉涼時重新抽發枝條。雖然枸杞不算是主流蔬菜，但它管理容易，病蟲害又不多，也是一個不錯的選擇。

Matrimony vine can grow for a few years. It wilts under hot weather and new shoots emerge when it gets cooler. It is easy to grow and not prone to pest or disease. Although matrimony vine is not a major crop, it is good to be included in one's planting plan to enhance crop diversity.

8 月下旬 -10 月
Late August to October

處暑
End of Heat

中至深 (>15)
Medium to deep (>15)

60-70 日
60-70 days

20-30 厘米
20-30 cm

種植貼士 Growing tips

- 以堆肥、骨粉等混入田畦中作基肥，枝葉生長時，再追施氮肥(如麩粉)

- 以插枝方法繁殖。5、6 月收成完結時，順帶在枝條的中至底部份選取 10-12 厘米部份(最少帶 5-6 個節位，粗身但未木質化部份；其實，在街市購買的新鮮枸杞，只要狀態良好，也能用來插枝)。選取枝條後，綑縛一起浸在水中(水浸至枝條一半高度左右，上下方向不能倒轉)，待其發根後可移植泥中

- 也可將舊有枸杞植株的莖、葉整條割去，減低損耗，讓其渡過炎熱，待秋涼來臨時再發新梢

- 6-8 月天氣炎熱，可以蔭網為枸杞蓋頂

- 可採用免掘的方法種植

- 枸杞收成期由冬天到翌年的 5、6 月，收成時將整個枝條摘下。由於收成期長，因此中段期間要為作物補充肥料

- 雖然枸杞是多年生作物，但建議每隔一、兩年，便更換一次種植區域，一來避免連作，二來也防止老根過度累積，阻礙新枝條抽發

- Apply compost and bone meal as base fertilizer. Apply nitrogenous fertilizer (e.g. peanut cake) to support leaf growth.

- Propagate by a cutting. Prepare the cutting with 10-12cm bottom part of the stem at the end of the harvesting period, by June. The cutting should be thick but not woody, and bearing 5-6 nodes. Fresh vine from market can also be used for preparing cuttings. Place the cuttings upright in water and transplant them when roots develop.

- Propagation can also be done by removing all stems and leaves from the growing plant before the summer to reduce water loss and preserve nutrient. New shoots emerge in the following autumn.

- Cover the crop with a shade net from June to August when the weather is hot.

- The No-dig method can be used.

- The harvesting period starts in winter and lasts until June. Pick the whole shoot when harvesting. Replenish fertilizer at the mid-growing stage.

- Although it is perennial, the planting site should be changed every one or two years to avoid problems related to continuous cropping. Old roots also block growth of new shoots and they should be removed.

蔥 Spring Onion

常見品種主要是分蔥和水蔥。前者的香味較濃，但收成期較短；水蔥的生長條件則較為寬鬆。此外，外國也有「細香蔥」品種，植株細小但香味濃烈。

蔥是比較容易栽種的品種，用途廣泛，又能以盆栽種植，切合家居園圃的需要。不過，不同類別的蔥，其栽種時機及方法差異甚大，這點要小心處理。

Spring Onion and Water Onion are the two major varieties in China. The former has a shorter growing period and is stronger in flavour, while the growing requirement of the latter is less demanding. Chive is a variety of foreign origin, which is smaller in size and stronger in flavour.

Spring onion is easy to grow, is widely used and can be planted in a pot. It is suitable for growing in a home garden. The planting period and method varies among varieties.

分蔥（Spring onion）

- 8月下旬-9月
  Late August to September
- 處暑
  End of Heat
- 中（15-20）
  Medium (15-20)
- 40-50 日（蔥頭）
  40-50 (bulbs)
- 20-30 厘米
  20-30 cm

細香蔥（Chive）

- 8月下旬-9月
  Late August to September
- 處暑
  End of Heat
- 中（15-20）
  Medium (15-20)
- 60-70（種子）
  60-70 (seeds)
- 20-30 厘米
  20-30cm

水蔥（Water onion）

- 10月-4月
  October to April
- 寒露
  Cold Dew
- 中（15-20）
  Medium (15-20)
- 60-70（種子）
  60-70 (seeds)
- 20-30 厘米
  20-30 cm

水蔥
Water Onion

分蔥
Spring Onion

種植貼士 Growing tips

分蔥

- 街市售賣的乾蔥頭也能用來種植。種植時將乾蔥頭拆散，每一小粒作一單位，下種時小粒埋在畦內，尖位向天，僅露出田面，然後澆水，數天後便會長出嫩梢。種植後約2個月可收成
- 當蔥生長至有4、5束時，便能開始收成
- 收成時將不多於一半的蔥連根劇出來，過程需要小心，避免傷及保留的另外部份。這樣的話，保留的株植部份會延續抽發新葉
- 翌年4月天氣轉為濕暖時，分蔥的狀態會開始轉壞，這時應停止收成，讓枝條自由生長，6月時天氣酷熱，蔥的葉會枯萎，剩下基部。這時可將蔥的基部完整地劇起，風乾成為乾蔥頭，待秋播時再種植

水蔥

- 水蔥採用種子種植，生長條件寬鬆
- 除了炎夏最熱幾個月外，其他時間都能栽種

細香蔥

- 植株細小，較適合以盆栽種植

Spring onion

- A dry bulb cluster from the market can be used for propagation. Break the cluster and plant each bulb individually with its tip pointing upward. Young leaves emerge in a few days and it is ready for harvest in two-months.
- Harvest when the spring onions spread out into 4-5 bunches.
- Harvest by pulling and uprooting the plant. Leave half of the bulbs in the soil to allow new growth.
- Stop harvesting when the plant's condition gets worse under hot weather. When its leaves wilt in June, harvest the cluster of bulbs, dry it, and sow it in the following autumn.

Water onion

- Seed is used in propagation.
- Can be grown all year round except summer months.

Chive

- Small in plant size. Suitable for pot-planting.

芫荽 Coriander

華南地區常用香料。也有外國品種「番茜」，
但多用來伴碟，較少人栽種。

It is widely used as seasoning in Southern China.
Parsley is a variety of foreign origin. It is often used
as gamish and it is less popular in local farms.

 9月-10月
September to October

白露
White Dew

 淺（10-15）
Shallow (10-15)

 90-120日
90-120 days

 5-10厘米
5-10 cm

種植貼士
Growing tips

- 較少病蟲害
- 生長緩慢，小小一株芫荽需種植約100天
- 忌移植，這會令根部造成損傷，宜採露地播種
- 開花前收採

- Not prone to pest or disease.
- Grows slowly.
- Outdoor sowing. Avoid transplanting to prevent root damage.
- Harvest before flowering.

水果
Fruits

由於生果一般體積較大，要引入家居農圃之中，存在較多的限制。不過，如果我們能夠挑選一些小巧品種，配合適當容器（如瓦缸）及一些修剪工作，還是有一定發揮空間的。

以下是幾款較容易引入都市環境的生果品種。

The choice of fruit trees for home gardening is limited by space constraints. Petite varieties can be planted in a proper container (e.g. a tile pot) with regular pruning. This section lists some fruits that are suitable for growing in a home garden.

菠蘿 Pineapple

菠蘿曾是香港的主要水果，適合於坡地種植。時至今日，香港不少地名仍然是「菠蘿」為名。由於菠蘿的個子小，一個花盆已能栽種，故很適合於家居環境(尤其是天台)中栽種。

Pineapple used to be a major fruit grown in Hong Kong and that's why several places carry 'Por-law (the Chinese name of pineapple) in their names. It is suitable for planting on slopes and pot-planting in a home garden (especially in a roof-top garden).

種植貼士 Growing tips

- 菠蘿的側芽及頂芽(見小圖)，都可用來繁殖。芽根越「老身」，存活率越高。一般來說，由芽定植至首次收成，大概要2-3年時間。其後，植株每年可結果一次。

- 於街市購買的新鮮菠蘿，也可以嘗試用其頂芽作繁殖，但應選擇較「老身」的頂芽。方法是先將芽處於陰涼地方2-3星期，待其氣根發展；然後直接定植於田畦或盆栽泥土表面

- 4月及9月是施肥的良好時機

- 主要收成期是7、8月，果實轉呈黃色時便可收採

- 果實位於植株頂部，收割後，待根部抽新側芽；側芽長出後，便可割去舊的果梗部份。新的側芽翌年會再度開花結果

- 建議種植的花盆不少於24cm直徑

- Propagated by bud of sucker and crown. A mature bud has a higher success rate. It takes 2-3 years for fruit to be produced in the first time and thereafter fruit will be produced every year.

- The crown of pineapple bought from a market can be used for propagation. Place the crown in a shady and airy place for 2-3 weeks. Transplant into soil when aerial roots develop.

- April and September are the best times to apply fertilizer.

- Harvesting period is July and August. Harvest when the fruit starts turning yellow in colour. The fruit at the top of the plant is harvested.

- Lateral nodes develop after harvesting. New nodes will flower and bear fruit the following year.

- Pot with a diameter wider than 24cm is needed for pot planting.

檸檬 Lemon

柑桔類乃水果家族中的主要成員，其中檸檬的植株較小，較容易引入家居農圃之中。品種方面，有西檸、唐檸之分，前者較為普遍。近年，也流行栽種泰國青檸。

檸檬四季均會開花，很適合小型農圃的需要。其枝幹帶有硬尖刺，編排種植時要加倍留意。

Lemon is a citric plant of small size, which is suitable for home-growing. Both Western and Chinese Lemon are available in the market while the former is more popular.
Kaffir lime is also getting popular. Lemon flowers all year around. Beware of the spikes on the stems when arranging the location of the plant in the garden.

種植貼士 Growing tips

- 春、夏、秋季均會長新梢，為免枝條混亂及控制其吋討（盆栽種植的檸檬不宜高於 1.5 米），需定期進行修剪，除去禿枝及過密枝條；並為過長的枝幹進行短截，避免枝條老化

- 如果是盆栽種植，每隔一、兩年將植株從盆中倒出，剪去死根及添施新的肥料，這樣能有助延長植株的壽命

- 果實轉黃色時便可收採

- New shoots emerge all year except in winter. For the plant size to fit a household garden, regular pruning is needed to remove bare branches. The optimum size is below 1.5 m.

- Every one or two years take the plant out of the pot to remove dead roots and apply fertilizer. This helps extend the life of the plant.

- Harvest when the fruit starts turning yellow.

熱情果 Passion Fruit

熱情果是理想的常綠多年生攀援植物。
Passion Fruit is a good evergreen perennial climber.

熱情果屬多年生常綠攀援植物。可與蔭棚的設計
相融合。由於其果肉與種子糾纏一起，鮮食不太
方便；但用於沖泡飲品，則非常合適。

Passion fruit is a perennial, evergreen climber which
can be used for setting up a shade shelter. Its seeds are
embedded in the pulp and it is best used in tea.

種植貼士
Growing tips

- 熱情果的枝葉繁密，單株已足可長滿一個5平
  方米棚架。為讓根部有足夠生長空間，建議在
  棚架下設花槽

- 溫暖天氣下生長迅速，蔓未上棚頂時，可剪除
  側枝讓主幹粗壯一些。蔓生至棚頂後，便可讓
  其順着蔭棚空間自由成長

- 熱情果一年有兩次花期。每次收成後宜進行修
  剪及施肥，促助往後的產量

- 果實轉呈黃色便能收採

- Passion fruit grows vigorously under warm
  weather and one plant can occupy a 5m² trellis
  fully. Set up a planter with a trellis to provide
  sufficient space for the plant to grow.

- At the early growing stage, remove lateral
  branches to stimulate growth of the main stem.
  After the main shot reaches the top of the trellis,
  allow the plant to grow freely.

- It flowers twice per year. The flowers turn
  into fruit. Pruning and applying fertilizer after
  harvesting enhance productivity.

- Harvest when the fruit starts turning yellow in
  colour.

木瓜 Papaya

木瓜是熱帶常綠植物。由於果實味美、產量豐富、生長期短,很受農夫及園藝種植者鍾愛。定植後,不到一年時間便會進入收成期。木瓜的豐收期大致能維持1-1.5年,隨着植株越高,果實會變得越來越細。當產量明顯下降時,便應考慮更換新的植株及輪種地點。

Papaya is a tropical, evergreen plant. Its tasty fruit, high productivity and short growing period make it a popular choice for farmers and gardeners. A harvest will be ready within one year after transplanting.
The peak harvest period of the crop can be maintained for 1 to 1.5 years. After that, the plant gets taller but with smaller fruits. Change the planting site once obvious decrease in harvest is observed. Crop rotation should be adopted.

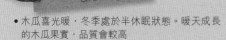

種植貼士
Growing tips

- 木瓜喜光暖,冬季處於半休眠狀態。暖天成長的木瓜果實,品質會較高
- 繁殖方面,木瓜多以種子育苗,播植期為2-3月或8-9月
- 果實轉呈橙色時便可收採
- 現時基因改造木瓜已廣泛在銷售市場出現,種植者應先核實栽種的是無基改品種,亦不要使用在生果鋪內購買的木瓜果實種子

- Papaya grows well under sunny and warm weather, and goes semi-dormant in winter. Fruits are more numerous if it is grown in a warm climate.
- Often propagated by nursery. Sowing either in February to March or August to September.
- Harvest when the fruit starts turning orange in colour.
- In view of the wide spread of genetically modified (GM) papaya in market, purchase non-GM seeds or seedlings from a reliable supplier.

食用香草
Culinary Herbs

香草用於烹調，已有數以千年的歷史。本地農場及花園常見的香草品種超過三十款，各具不同的用途。

品種

現時香港栽種的食用香草，源自不同的國家，栽種的條件亦不盡相同。因此，我們在種植前，必需先瞭解清楚不同地區香草的生長特性。

1）地中海區域的香草

包括迷迭香、百里香、鼠尾草、薰衣草、小白菊及馬祖林等。地中海香草原本生長於溫暖氣候和乾燥土壤，因此不喜歡香港夏季潮濕、容易積水的環境。種植時採用較能疏水的泥土，能有效改善栽種的條件。

2）熱帶香草

包括香茅及羅勒。香港位於亞熱帶，除冬季外，大多數時間它們都能夠生長良好。遇上既濕且冷的天氣時，盆栽種植者可把香草搬移至較為和暖的地方。

3）中國的香草

包括蔥、蒜、紫蘇等。這些品種多已適應本地氣候，只要在適合的時候栽種，要成功收成並不困難。

Herbs have been widely used in cooking, beverages and medicine for thousands of years. There are over 30 varieties of culinary herbs planted widely in local Hong Kong farms and gardens for various purposes.

Varieties

Culinary herb plants that originate from other countries may require growing conditions that are different from our local climate. Therefore, it is essential to know their origins and growing requirements before planting.

1) Mediterranean herbs

Rosemary, thyme, sage, lavender, feverfew and marjoram are representative examples of Mediterranean herbs. Most Mediterranean herb plants thrive in that region's warm temperature and dry soil. They do not favour the high humidity and seasonal very wet soil in Hong Kong, therefore one will have the most success with a good quality well-draining potting soil.

2) Tropical herbs

Lemon grass and basil are representative examples of tropical herbs. Tropical herbs generally grow well under the sub-tropical climate in Hong Kong, except in winter. During cold and wet days there is a need to move potted tropical herb plants to warmer locations.

3) Chinese herbs

Garlic and perilla are examples of common Chinese culinary herbs. They adapt well to the local climate. Planting at the right time normally yields a good harvest.

百里香 Thyme

葉片用作料理、沖茶、製香包。
鋪地而生的根莖植物，喜肥沃、排水良好的土壤，需充足日照。

Leaves are for culinary use and tea-making.
Grows as ground cover. Thrives in fertile soil with good drainage and full sun.

9-11月
September to November

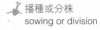
播種或分株
sowing or division

120-150天
120-150 days

茴香 Fennel

9-12月
September to December

播種
sowing

90-100天
90-100 days

莖和葉可用於料理，尤其適用於魚及烤肉，種子可用沖茶或製作麵包。
喜濕潤、排水良好的肥沃土壤，需充足日照。

Leaves and stems are for culinary use (perfect with fish and roast meats); seeds can be used in making tea or bread.
Likes humid and fertile soil with good drainage, needs sufficient daylight.

蒔蘿 Dill

葉、花、種子均可用於料理，尤其適用
於魚類。

形態、種植方法類似茴香。

Leaves, flowers and seeds are for culinary use
(perfect with fish).

Shape and sowing method are similar to fennel.

🔍 9-12 月
September to December

🌱 播種
sowing

🌼 50天收葉，100天收種子
50 days collecting leaves, 100 days collecting seeds.

番茜 Parsley

葉及莖可作沙律、湯及點綴其他料理。

喜歡排水良好的土壤，全日照及半日照環境均可種植

Leaves and stems can be used in salad, soup and other dishes. Often
used as garnish.

Likes fertile soil with good drainage, full or half-day sunlight.

🔍 9-12 月
September to December

🌱 播種
sowing

🌼 80-100天
80-100 days

鼠尾草 Sage

📅 9-11月
September to November

🌱 播種
sowing

🌿 100-120天
100-120 days

葉片用作料理、沖茶，具良好
的提神作用。
喜歡全日照及排水良好的土壤。

Leaves can be used in dishes or
tea-making. It is known to be a good
brain stimulant to enhance one's
alertness.
Grow in full sun and well-drained soil.

薄荷 Mint

薄荷品種繁多，最常見的是綠薄荷。
葉及莖可作香草茶及料理。
多年生蔓生植物，喜濕潤和肥沃的土壤，全日照或半日
照的環境中均可種植。

Leaves and stems can be used in making herbal tea and dishes.
There are many variations of mint; spearmint is the most common
variety.
Perennial trailing plant, likes humid and fertile soil, full or half-day
sunlight.

📅 3-5月，9-11月
March to May, September to
November

🌱 播種、插枝，分株
sowing, cutting and division

🌿 50天（插枝）
50 days (cutting)

香蜂草 Lemon Balm

葉及莖可作香草茶及料理。
種植方法類似綠薄荷。

Leaves and stems can be used in making herbal tea and dishes.
Planting method is similar to that of spearmint.

3-4月, 9-10月
March to April, September to October

播種、插枝、分株
sowing, cutting and division

60天（插枝），100天（種子）
60 days (cutting), 100 days (seed)

皇家便士 Pennyroyal

可作藥用或沖茶。
喜涼，鋪地而生，全日照及半日照環境均可種植。

Medical use or tea-making.
Likes cool weather; grows as ground cover; full or half-day sunlight.

9-11月
September to November

播種或分株
sowing or division

90-100天
90-100 days

牛至 Oregano

葉片用作料理、沖茶。

喜歡全日照及排水良好的土壤，喜偏鹼的泥土。

Leaves are used in dishes and tea-making.
Grows in full sun and well-drained soil. Alkaline soil is favoured.

 9-11月
September to November

 播種
sowing

90-100天
90-100 days

琉璃苣 Borage

9-11月
September to November

 播種
sowing

80-100天
80-100 days

葉片及花用作料理或飲品。

喜歡全日照及排水良好的土壤，長大後體積甚大，需預留足夠空間。

Leaves and flowers are used in dishes and beverages.
Grows in full sun and well-drained soil. Very big in size when fully grown, hence reservation sufficient space is necessary.

迷迭香 Rosemary

9-11 月
September to November

播種，插枝
sowing, cutting

100-120天
100-120 days

葉片用作料理、沖茶。
喜歡全日照、乾爽氣候及排水良好的土壤。
Leaves are used in cooking and tea-making.
Grows in full sun, dry weather and well-drained soil.

酸模 Sorrel

9-11 月
September to November

播種
sowing

80-100天
80-100 days

葉片可用於料理及湯。
多年生植物。喜歡肥沃、排水良好的土壤。
Leaves can be used in dishes and soup.
Perennial, likes fertile soil with good drainage.

羅勒 Basil

2-5月
February to May

播種或插枝
sowing or cutting

70-80天
70-80 days

葉片可作調味料及香草茶，亦可驅走蚊蟲。
一年或多年生植物，品種甚廣，花及葉的顏色變化很大。喜排水良好和疏鬆的土壤，在全日照及部份遮蔭的環境中均可種植。忌低溫(10度以下)。

Leaves are used as seasoning, to make herbal tea and to repel mosquitoes and other insects.
Annual or perennial plant with many varieties; colours of flowers and leaves vary a lot; likes loose soil with good drainage, full or half-day sunlight; avoid low temperayure (below 10°C).

香茅 Lemon Grass

3-5月, 9-10月
March to May, September to October

分株
division

100-120天
100-120 days

葉及莖可作沖茶及料理。
多年生，喜歡溫暖氣候，對土壤要求不高，容易打理。

Leaves and stems can be used in making tea and dishes.
Perennial, likes a warm climate, adaptable to different types of soil, easy to take care.

芸香 / 臭草 Rue

 3-5月, 9-11月
March to May, September to November

 插枝
Cutting

60天
60 days

可用作料理,常用於綠豆沙。此外,臭草亦有驅蟲功效。

多年生,有黃色小花。

喜排水良好和疏鬆的土壤,需充足日照。

Perfect in Chinese mungbean soup. Repelling insects.
Perennial, small yellow flowers.
Likes loose soil with good drainage, needs sufficient daylight.

紫蘇 Perilla

3-5月
March to May

播種或插枝
sowing or cutting

70-80天
70-80 days

葉片可作料理、沖茶。

喜歡排水良好的土壤,全日照及半日照環境均可種植。

Leaves are used in cooking and tea-making.
Likes well-drained soil; full or half-day sunlight.

金蓮花 **Nasturitium**

葉及花可作料理。
喜歡全日照及排水良好的土壤。

Leaves and flowers are used in cooking.
Grow best in full sun and soil with good drainage.

 10-12月
October to December

 播種
sowing

 70-90天
70-90 days

金盞菊 **Calendula**

花瓣可作料理。
喜歡全日照及排水良
好土壤。

Petals are used in culinary.
Grows well under full sun
on well-drained soil.

 10-12月
October to December

 播種
sowing

90-120天
90-120 days

Chapter

05

有機十問
Ten Questions
about Sustainable Farming

永續農業對我們有多重要？
Why is sustainable farming important for us?

近數十年地球上出現的種種環境問題，包括泥土侵蝕、生物多樣性減少、生態環境遭破壞、污染及氣候轉變等，都與慣行農業的模式有着密切關係。要逆轉這災難性的趨勢，便需要在全球角度推動永續農業，例如永續栽培（樸門）、生物動力農法、自然農法等等。為了未來的人類及自然環境，我們急切需將現有的糧食系統，變得更具可持續性。

Conventional agriculture has been one of the largest drivers of soil erosion, biodiversity loss, habitat destruction, pollution and climate change in the last few decades. To reverse this devastating trend, the sustainable agriculture movement in the forms of permaculture, biodynamic farming, natural farming and organic farming have been evolving worldwide. There is an urgency for a global transition towards a sustainable food system globally, so as to ensure a more sustainable future for people and nature.

Q2
消費者有甚麼方法，支持更符合永續原則的食物系統？
How can a consumer support growth of a more sustainable food system?

要支持永續農業，我們可以：
* 選購最短運輸距離、符合公平貿易原則的有機食品。
* 透過參與「社區支持農業」計劃（CSA）、購買農墟產品，支持本地農業。
* 支持本地的小商戶及食物生產者。
* 不時不食，維護傳統飲食文化。
* 自己也種植一些食物吧！

To support transition to sustainable agriculture, please:
* Choose fairly traded organic food which has travelled the least distance
* Support local agriculture by joining a Community Supported Agriculture (CSA) scheme and/or purchase food at a local Farmers' Market
* Support small local food stores and food production initiatives
* Eat seasonally and preserve traditional food culture and cuisine
* Grow some food.

Q3 氣候轉變如何影響農業？
How does climate change affect agriculture?

　　農業與當區的氣候環境息息相關，氣候轉變對糧食生產帶來莫大的影響。近年，颱風、水災、乾旱及熱浪等極端自然災害愈益頻繁，大大加深了人類糧食安全的威脅。

Agriculture is highly dependent on specific climate conditions and climate change has started to significantly affect agriculture. The frequency and intensity of extreme weather events such as cyclones, floods, droughts and heat waves attributed to changing climatic conditions have increased in recent years and pose severe risks to food security across the world.

Q4 「石油頂峰」如何影響到農業及香港市民？
How does Peak Oil relate to agriculture and Hongkonger?

　　現今慣行農業及工業化的食物系統，包括糧食生產、貿易、分配等部份，均完全地依靠廉價的化石燃料來維持。香港超過九成的食物，進口自大概150個國家，糧食價格與燃油成本有着密切關係。對於「石油頂峰」（全球石油產量逐步下降，廉價石油的日子將會停止）的影響，我們現在還未有適當的應對方法，這導致問題在未來的日子會變得更嚴峻。

Convention agriculture and the industrialized food system are highly dependent on cheap fossil fuels in all aspects of food production, trading and distribution. Today, over 90% of the food consumed in Hong Kong is imported from around 150 countries – the price of our food is closely linked to the price of fossil fuels. We are currently unprepared for the effects of Peak Oil (the gradual reduction in global oil production and the end of cheap oil), which will intensify in future.

Q5 甚麼是永續栽培（又稱「樸門」）？
What is Permaculture?

　　永續栽培是按照生態原則而演繹的設計方法。設計以瞭解並依從自然規律為法則，而並非與自然對抗。它以和諧協調的方向來整合環境與地貌，以永續方式提供人類物質或非物質的需要，包括食物、能源及棲息居所等。對於現今社會存在的能源短絀、物種威脅等問題，永續栽培均以創意而務實的方法來應對。

　　現時，香港有不同的組織定期開辦「永續栽培設計證書」課程。

Permaculture is defined as a design system based on ecological principles, which work with, rather than against nature. It is a harmonious integration of landscape with people; providing their food, energy, shelter and other material and non-material needs in a sustainable way. Permaculture offers a practical, creative approach to the problems of diminishing resources and threats to our life support systems, which are now facing the world. Permaculture Design Certificate courses are organized regularly by various organizations in Hong Kong.

Q6 我在哪裏可以購買到本地農夫的有機產品？
Where can I buy organic produce from local farmers?

有機農墟，是消費者支持本地有機農夫的最好場所。

早年的有機農墟，僅屬個別的短期活動；但時至今日，太和、中環天星碼頭、美孚等地方，均已有定期的農墟，在這裏，消費者可以直接向本地農戶購買新鮮的有機產品。農墟同時也是一個公平交易的地方，消費者可以知道他們吃的東西來自甚麼地方；而農夫也能在不受剝削的情況，更公平地取得應有回報。

Organic farmers' markets are the best places for consumers to express their support and gratitude to local organic farmers. Starting from occasional events, farmers markets are now held regularly at Tai Wo, Central Star Ferry Pier 7, Mei Foo and other locations in Hong Kong. At farmers' markets, consumers can meet and purchase freshly harvested organic produce directly from local farmers. It is a fair trade platform where consumers know exactly where the products come from while farmers enjoy a fairer return for their skills and physical effort.

Q7 農業與香港「無關」？
Is agriculture relevant to Hong Kong?

香港七百萬人的糧食供應，從來都是一個嚴肅的問題。受到石油頂峰、氣候轉變的影響，食物供應的問題未來將會變得更具挑戰。為了保障糧食安全及加強公眾關注，我們需設法維持香港現存的農地，並積極開展香港的都市農業。

It is important for us to realize that 'food security' for the population of over seven million in Hong Kong is a very serious issue and will be more challenging under the escalating effects of Peak Oil and climate change. In order to safeguard and increase public interest in this important issue, it is vital that we prevent further loss of our existing farmland and greatly expand urban agriculture in an urgent and creative manner.

Q8 都市農業，在香港可行嗎？
Is urban agriculture feasible in Hong Kong?

除了保護農地，發展都市農業同樣需要更多公眾對本土食物生產的支持及參與。栽種食物有時並非想像般複雜，誠然，我們面對空間不足及環境方面的種種限制。但許多情況下，我們仍能夠在家居中種植一些食物。

Apart from preserving existing farmland, expanding urban agriculture requires a much wider public participation and support of local food production. Growing some of your own food is a lot less complicated than people might think. Shortage of space and environmental constraints can be challenging, but it is usually possible to start by grow some food at home.

隨着世界各地的有機產品需求不斷上升，越來越多消費者透過超級市場等渠道購買有機食物，他們需要第三者監察來確認其可靠性。各個認證機構會按其標準，為不同的農產品提供有機檢測認證服務。

不過，並非所有消費者都需要有機認證的。消費者可透過農墟、社區支持農業（CSA）等平台，直接與當地有機農民接觸。這些實質的溝通，比起認證其實更為實在。本園亦沒有申請有機認證，我們與消費者的關係是以信任為基礎。

As more and more consumers purchase organic products via channels like supermarkets, organic certification systems run by third-parties have evolved to address the growing global market in quality assurance. Individual certification bodies are operating with their own service marks and certification standards.

Organic certification is regarded as unnecessary by some consumers in farmers markets or collaborative platforms like Community Supported Agriculture (CSA) where they have a direct bond with local organic farmers, which is more tangible than a label. For instance, Kadoorie Farm and Botanic Garden have never sought certification and our relationship with consumers is trust-based.

　　我們的都市，需要轉化成一個更本土化、符合永續食物系統原則的社會。而這個系統，應該以小規模、低度能源消耗的有機農場，以及社區為本的都市農圃作為基礎。除了提供食物，都市農圃還可創造一個多功能的庭園，促進保育、康樂、社區建設、水及有機資源循環等作用。

　　天台、露台及公園，都有潛力成為栽種食物的地方。現今，農地只佔有香港6.1%的空間，面積與6,800個足球場相約。今日開始，我們便要展開行動，保育它們免遭發展吞噬，為提高本地糧食供應而努力！

We have to transit towards a more localized and sustainable food system that is built upon small, less energy-intensive organic farms and community-based urban gardens. Apart from food supply, urban gardens will help us create a multi-functioning landscape that serves nature conservation, recreation, community-building, water and organic resource recovery.

Roof tops, balconies and parks are other potential places for growing our own food. Agricultural land covers only 6.1% of Hong Kong and is equivalent to the size of 6,800 football fields - we must act now to safeguard it from development, to assure some level of food security!

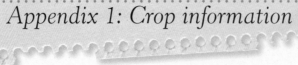

附錄一：農作物資料
Appendix 1: Crop information

葉菜類
Leaf Vegetables

中英名 Common Name	學名 Scientific name	分類 Family	種植月份 Month of growing	
生菜 Lettuce	*Lactuca sativa*	菊科 Asteraceae	8月下旬至翌年2月 Late Aug - Feb	
油麥菜 Indian Lettuce	*Lactuca sativa*	菊科 Asteraceae	8月下旬至翌年2月 Late Aug - Feb	
茼蒿 Garland Chrysanthemum	*Chrysanthemum coronarium*	菊科 Asteraceae	8月下旬至10月 Late Aug - Oct	
菠菜 Spinach	*Spinacia oleracea*	藜科 Chenopodiaceae	9月至翌年2月 Sep - Feb	
白菜 White Cabbage	*Brassica chinensis*	十字花科 Brassicaceae	4月至10月（視乎品種） Apr - Oct (depending on variety)	
菜心 Flowering Cabbage	*Brassica parachinensis*	十字花科 Brassicaceae	4月至10月（視乎品種） Apr - Oct (depending on variety)	
芥蘭 Chinese Kale	*Brassica oleracea var. alboglabra*	十字花科 Brassicaceae	8月下旬至12月 Late Aug - Dec	
莧菜 Amaranth, Chinese Spinach	*Amaranthus tricolor*	莧科 Amaranthaceae	3月至9月 Mar - Sep	
潺菜 Basella	*Basella rubra*	落葵科 Basellaceae	3月至4月 Mar - Apr	
芹菜 Celery	*Apium graveolens var. dulce*	傘形科 Apiaceae	9月至10月 Sep - Oct	
椰菜 Cabbage	*Brassica oleracea*	十字花科 Brassicaceae	8月下旬至12月 Late Aug - Dec	
黃芽白 Chinese Cabbage	*Brassica rapa*	十字花科 Brassicaceae	8月下旬至9月 Late Aug - Sep	

株行距(cm) **Planting Distance**	播種方法 **Propagation Method**	生長期（日） **Growing Period (day)**
20-25	培苗 Nursery	60-70
25-30	培苗 Nursery	70-80
15-20	行播、撒播 Sowing in line, Broadcast sowing	50-60
15-20	撒播 Broadcast sowing	60-70
10-20	撒播、培苗 Broadcast sowing, Nursery	40-70
8-20	撒播、培苗 Broadcast sowing, Nursery	40-70
15-25	培苗 Nursery	60-80
5-8	撒播 Broadcast sowing	30-40
行距：30 Between rows：30 株距：15-20 Between plants：15-20	行播 Sowing in line	50-60
30-40	培苗、穴播 Nursery, Sowing in hole	120-150
40-50	培苗 Nursery	90-120
30-40	培苗 Nursery	70-90

中英名 Common Name	學名 Scientific name	分類 Family	種植月份 Month of growing	
芥菜 Mustard	*Brassica juncea*	十字花科 Brassicaceae	4月至10月（視乎品種） Apr - Oct (depending on variety)	
通菜 Water Spinach	*Ipomoea aquatica*	旋花科 Convolvulaceae	3月至4月 Mar - Apr	
君達菜 Swiss Chard	*Beta vulgaris*	藜科 Chenopodiaceae	9月至10月 Sep - Oct	
西洋菜 Watercress	*Nasturtium officinale*	十字花科 Brassicaceae	8月下旬至9月 Late Aug - Sep	
韭菜 Chinese Chive	*Allium tuberosum*	石蒜科 Amaryllidaceae	3月至4月，9月 Mar - Apr, Sep	

瓜類
Gourds

中英名 Common Name	學名 Scientific name	分類 Family	種植月份 Month of growing	
青瓜 Cucumber	*Cucumis sativus*	葫蘆科 Cucurbitaceae	2月下旬至8月 Late Feb - Aug	
節瓜 Hairy Gourd	*Benincasa hipida var. chieh-qua*	葫蘆科 Cucurbitaceae	2月下旬至5月 Late Feb - May	
冬瓜 Wax Gourd	*Benincasa hispida*	葫蘆科 Cucurbitaceae	2月下旬至4月 Late Feb - Apr	
絲瓜 Silky Gourd	*Luffa acutangula*	葫蘆科 Cucurbitaceae	2月下旬至5月 Late Feb - May	
水瓜 Sponge Gourd	*Luffa cylindrica*	葫蘆科 Cucurbitaceae	2月下旬至4月 Late Feb - Apr	
苦瓜 Bitter Cucumber	*Momordica charantia*	葫蘆科 Cucurbitaceae	2月下旬至5月 Late Feb - May	
葫蘆瓜 Bottle Gourd	*Lagenaria siceraria*	葫蘆科 Cucurbitaceae	2月下旬至4月 Late Feb - Apr	
南瓜 Pumpkin	*Cucurbita spp.*	葫蘆科 Cucurbitaceae	2月下旬至4月 Late Feb - Apr	
佛手瓜 Chayote	*Sechium edule*	葫蘆科 Cucurbitaceae	2月下旬至9月 Late Feb - Sep	
翠玉瓜 Zucchini	*Cucurbita pepo*	葫蘆科 Cucurbitaceae	8月下旬至10月 Late Aug - Oct	

株行距（cm） **Planting Distance**	播種方法 **Propagation Method**	生長期（日） **Growing Period (day)**
10-40	播種、培苗 Sowing, Nursery	40-80
行距：30 Between rows：30 株距：15-20 Between plants：15-20	行播、扦插 Sowing in line, Cutting	50-60
40-50	穴播、培苗 Sowing in hole, Nursery	70-90
20-30	分株、扦插 Division, Cutting	50-70
20-30	播種、分株 Sowing, Division	50-70（分株） 50-70 (Division)

株行距（cm） **Planting Distance**	播種方法 **Propagation Method**	生長期（日） **Growing Period (day)**
40-50	培苗 Nursery	50-70
60	培苗 Nursery	60-70
60-90	培苗 Nursery	90-120
60	培苗 Nursery	60-70
60-90	培苗 Nursery	60-70
45-60	培苗 Nursery	60-70
60-90	培苗 Nursery	60-70
150-200	培苗 Nursery	90-120
150-200	成熟的果實 Germinated by mature gourd	100-150
45-60	培苗 Nursery	60-70

豆類
Beans and Peas

中英名 Common Name	學名 Scientific name	分類 Family	種植月份 Month of growing	
豆角 String Bean	*Vigna sesquipedalis*	蝶形花科 Fabaceae	2月下旬至6月 Late Feb - Jun	
玉豆 Snap Bean	*Phaseolus vulgaris*	蝶形花科 Fabaceae	2月下旬至9月 Late Feb - Sep	
綠豆 Mung Bean	*Phaseolus aureus*	蝶形花科 Fabaceae	2月下旬至9月 Late Feb - Sep	
荷蘭豆 Sugar pea	*Pisum sativum*	蝶形花科 Fabaceae	9月下旬至10月 Late Sep - Oct	
蜜糖豆 Honey pea	*Pisum sativum*	蝶形花科 Fabaceae	9月下旬至10月 Late Sep - Oct	

茄科
Solanaceae

中英名 Common Name	學名 Scientific name	分類 Family	種植月份 Month of growing	
番茄 Tomato	*Lycopersicon esculenta*	茄科 Solanaceae	8月下旬至10月 Late Aug - Oct	
茄子 Eggplant	*Solanum melongena*	茄科 Solanaceae	春播：2月下旬至4月 Spring: Late Feb - Apr 秋播：8月下旬至9月 Autumn: Late Aug - Sep	
甜椒 Bell pepper	*Capsicum frutescens var. grossum*	茄科 Solanaceae	2月下旬至4月 Late Feb - Apr	
辣椒 Hot Pepper	*Capsicum frutescens*	茄科 Solanaceae	春播：2月下旬至4月 Spring: Late Feb - Apr 秋播：8月下旬至9月 Autumn: Late Aug - Sep	

株行距（cm） **Planting Distance**	播種方法 **Propagation Method**	生長期（日） **Growing Period (day)**
40-50	培苗、穴播 Nursery, Sowing in hole	50-60
15-40	培苗、穴播 Nursery, Sowing in hole	30-60
10-20	撒播、行播 Broadcast sowing, Sowing in line	40-50（綠肥） 40-50 (For green manure) 70-80（收成） 70-80 (For harvest)
行距：40-50 Between rows: 40-50 株距：15-20 Between plants: 15-20	行播 Sowing in line	60-70
行距：40-50 Between rows: 40-50 株距：15-20 Between plants: 15-20	行播 Sowing in line	60-70

株行距（cm） **Planting Distance**	播種方法 **Propagation Method**	生長期（日） **Growing Period (day)**
30-60	培苗 Nursery	40-90
50-60	培苗 Nursery	80-90
40-50	培苗 Nursery	80-90
30-40	培苗 Nursery	100-120

根莖類
Root and Stem

中英名 **Common Name**	學名 **Scientific name**	分類 **Family**	種植月份 **Month of growing**	
白蘿蔔 White Radish	*Raphanus sativus*	十字花科 Brassicaceae	早水蘿蔔：8月下旬至9月 Early Radish: Late Aug - Sep 遲水蘿蔔：10月至11月 Late Radish: Oct - Nov	
甘筍 Carrot	*Daucus carota*	繖形花科 Apiaceae	9月至翌年2月 Sep - Feb	
蕪菁 Turnip	*Brassica rapa var. rapa*	十字花科 Brassicaceae	9月至12月 Sep - Dec	
紅菜頭 Beetroot	*Beta vulgaris*	藜科 Chenopodiaceae	9月至12月 Sep - Dec	
芥蘭頭 Kohlrabi	*Brassica oleracea var. gongylodes*	十字花科 Brassicaceae	8月下旬至12月 Late Aug - Dec	

雜糧
Field crops

中英名 **Common Name**	學名 **Scientific name**	分類 **Family**	種植月份 **Month of growing**	
番薯 Sweet Potato	*Ipomoea batatas*	旋花科 Convolvulaceae	3月至4月；8月至9月 Mar - Apr; Ang - Sep	
木薯 Cassava	*Manihot esculenta*	大戟科 Euphorbiaceae	2月下旬至3月 Late Feb - Mar	
芋頭 Taro	*Colocasia esculenta*	天南星科 Araceae	2月下旬至3月 Late Feb - Mar	
薑 Ginger	*Zingiber officinale*	薑科 Zingiberaceae	2月下旬至3月 Late Feb - Mar	
沙葛 Yam Bean	*Pachyrhizus erosus*	蝶形花科 Fabaceae	2月下旬至6月 Late Feb - Jun	

株行距(cm) **Planting Distance**	播種方法 **Propagation Method**	生長期（日） **Growing Period (day)**
20-35	行播 Sowing in line	70-110
5-20	行播、撒播 Sowing in line, Broadcast sowing	60-70
20-30	培苗、行播 Nursery, Sowing in line	70-90
15-20	培苗、穴播 Nursery, Sowing in hole	120
30-40	培苗 Nursery	70-80

株行距(cm) **Planting Distance**	播種方法 **Propagation Method**	生長期（日） **Growing Period (day)**
30-40	扦插 Cutting	180-240
50-60	扦插 Cutting	240-280
50-60	根部繁殖 Root division	210-240
行距：30-40 Between rows: 30-40 株距：20-30 Between plants: 20-30	根部繁殖 Root division	150-180（子薑） 150-180 (For young ginger) 240-270（老薑） 240-270 (For old ginger)
20 - 30	行播、穴播 Sowing in line, Sowing in hole	100-150

其他蔬菜
Other vegetables

中英名 Common Name	學名 Scientific name	分類 Family	種植月份 Month of growing	
椰菜花 Cauliflower	*Brassica oleracea var. botrylis*	十字花科 Brassicaceae	8月下旬至12月 Late Aug - Dec	
西蘭花 Broccoli	*Brassica oleracea var. italica*	十字花科 Brassicaceae	8月下旬至12月 Late Aug - Dec	
粟米 Corn	*Zea mays*	禾本科 Poaceae	2月下旬至4月、8月下旬至9月 Late Feb - Apr; Late Aug - Sep	
秋葵 Okra	*Hibiscus esculentus*	錦葵科 Malvaceae	2月下旬至4月 Late Feb - Apr	
枸杞 Matrimony vine	*Lycium chinense*	茄科 Solanaceae	8月下旬至10月 Late Aug - Oct	
蔥 Spring Onion	*Allium cepa var. aggregatum*	石蒜科 Amaryllidaceae	分蔥：8月下旬至9月 Spring Onion: Late Aug - Sep; 水蔥：10月-4月 Water Onion: Oct-Apr 細香蔥：8月下旬至9月 Chive: Late Aug - Sep	
芫荽 Coriander	*Coriandrum sativum*	傘形科 Apiaceae	9月至10月 Sep - Oct	

株行距(cm) **Planting Distance**	播種方法 **Propagation Method**	生長期（日） **Growing Period (day)**
50-60	培苗 Nursery	80-120
40-50	培苗 Nursery	70-100
30-40	培苗、行播 Nursery, Sowing in line	70-80
40-50	穴播、培苗 Sowing in hole, Nursery	70-80
20-30	扦插 Cutting	60-70
20-30	根部繁殖(分蔥) Root Division (Spring Onion); 行播(水蔥) Sowing in line (Water Onion) 細香蔥	60-70（種子） 60-70 (Seed) 40-50（根部繁殖） 40-50 (Root division)
5-10	撒播 Broadcast sowing	90-120

生果
Fruits

中英名 Common Name	學名 Scientific name	分類 Family	
黃梨 Sand Pear	*Pyrus pyrifolia*	薔薇科 Rosaceae	
龍眼 Longan	*Dimocarpus longan Lour.*	無患子科 Sapindaceae	
荔枝 Lychee	*Litchi chinensis Sonn.*	無患子科 Sapindaceae	
橙 Orange	*Citrus sinensis*	芸香科 RutaceaeRutaceae	
柑桔 Tangerine	*Citrus reticulat*	芸香科 RutaceaeRutaceae	
柑 Mandarin	*Citrus reticulate cv. ponkan*	芸香科 RutaceaeRutaceae	
檸檬 Lemon	*Citrus limon*	芸香科 RutaceaeRutaceae	
木瓜 Papaya	*Carica papaya L.*	番木瓜科 Caricaceae	
柿 Persimmon	*Diospyros kaki Thunb.*	柿樹科 Ebenaceae	
菠蘿 Pineapple	*Ananas comosus (L.) Merr.*	鳳梨科 Bromeliaceae	
芒果 Mango	*Mangifera indica L.*	漆樹科 Anacardiaceae	
枇杷 Loquat	*Eriobotrya japonica (Thunb.) Lindl.*	薔薇科 Rosaceae	
楊桃 Carambola	*Averrhoa carambola L.*	酢漿草科 Oxalidaceae	
蕉 Banana	*Musa x paradisiaca L.*	芭蕉科 Musaceae	
番石榴 Guava	*Psidium guajava L.*	桃金孃科 Myrtaceae	
熱情果 Passion Fruit	*Passiflora delete sims cv. Flaviearpa*	西番蓮科 Passifloraceae	

生長期（年） Growing Period (Year)	花期（月） Month of Flowering	收成期（月） Month of Harvest
4-5（嫁接苗） 4-5 (Grafting)	2月至3月 Feb - Mar	7月至8月 Jul - Aug
4-5（嫁接苗） 4-5 (Grafting)	3月至4月 Mar - Apr	7月至8月 Jul - Aug
5-6（嫁接苗） 5-6 (Grafting)	2月至3月 Feb - Mar	6月至7月 Jun - Jul
3-4（嫁接苗） 3-4 (Grafting)	2月至3月 Feb - Mar	11月至12月 Nov - Dec
3-4（嫁接苗） 3-4 (Grafting)	2月至3月 Feb - Mar	11月至12月 Nov - Dec
3-4（嫁接苗） 3-4 (Grafting)	2月至3月 Feb - Mar	11月至12月 Nov - Dec
2-3（嫁接苗） 2-3 (Grafting)	全年 Whole yeawr	全年 Whole year
1（種子） 1 (Seed)	全年 Whole year	全年（9月至12月大收） Whole year (Great harvest in Sep - Dec)
4-5（嫁接苗） 4-5 (Grafting)	3月至4月 Mar - Apr	9月至10月 Sep - Oct
2（側芽） 2(Sucker) 3（冠芽） 3 (Crown)	2月至3月 Feb - Mar	7月至8月 Jul - Aug
3（嫁接苗） 3 (Grafting)	3月至4月 Mar - Apr	7月至8月 Jul - Aug
3-4（嫁接苗） 3-4 (Grafting)	11月至12月 Nov - Dec	3月至4月 Mar - Apr
3-4（嫁接苗） 3-4 (Grafting)	全年 Whole year	全年（9月至10月大收） Whole year (Great harvest in Sep - Oct)
1（分芽） 1 (Division)	全年 Whole year	全年（6月至9月大收） Whole year (Great harvest in Jun - Sep)
1-2（圈枝苗） 1-2(Layering)	3月至5月、9月至10月 Mar- May; Sep-Oct	8月至11月 Aug - Nov
1-2（插枝苗） 1-2 (Cutting)	不穩定 Unstable	

香草
Herbs

中英名 Common Name	學名 Scientific name	分類 Family	
百里香 Thyme	*Thymus vulgaris*	唇形科 Lamiaceae	
茴香 Fennel	*Foeniculum vulgare*	傘形科 Apiaceae	
蒔蘿 Dill	*Anethum graveolens*	傘形科 Apiaceae	
番茜 Parsley	*Petroselinium crispum*	傘形科 Apiaceae	
鼠尾草 Sage	*Salvia officinalis*	唇形科 Lamiaceae	
薄荷 Mint	*Mentha species*	唇形科 Lamiaceae	
香蜂草 Lemon Balm	*Melissa officinalis*	唇形科 Lamiaceae	
皇家便士 Pennyroyal	*Mentha pulegium*	唇形科 Lamiaceae	
牛至 Oregano	*Origanum majorana*	唇形科 Lamiaceae	
迷迭香 Rosemary	*Rosmarinus officinalis*	唇形科 Lamiaceae	
酸模 Sorrel	*Rumex acetosa*	蓼科 Polygonaceae	
羅勒 Basil	*Ocumum basilicum*	唇形科 Lamiaceae	
香茅 Lemon Grass	*Cymbopogon citratus*	禾本科 Poaceae	
萬壽菊 Marigold	*Calendula officinalis*	菊科 Asteraceae	
芸香/臭草 Rue	*Ruta graveolens*	芸香科 Rutaceae	
紫蘇 Perilla	*Perilla frutescens*	唇形科 Lamiaceae	
琉璃苣 Borage	*Borago officinalis*	紫草科 Boraginaceae	
金蓮花 Nasturtium	*Tropaeolum majus*	旱金蓮科 Tropaeolaceae	
金盞菊 Calendula	*Calendula officinalis*	菊科 Asteraceae	
芳香萬壽菊 Mt. Lemmon Marigold	*Tagetes lemmonii*	菊科 Asteraceae	

種植月份 Month of growing	播種方法 Propagation Method	生長期（日） Growing Period (day)
9月至11月 Sep - Nov	播種、分株 Sowing, Division	120-150
9月至12月 Sep - Dec	播種 Sowing	90-100
9月至12月 Sep - Dec	播種 Sowing	50（收葉）、100（種子） 50 (For leaves), 100 (For seeds)
9月至12月 Sep - Dec	播種 Sowing	80-100
9月至11月 Sep - Nov	播種 Sowing	100-120
3月至5月、9月至11月 Mar - May; Sep - Nov	播種、插枝、分株 Sowing, Cutting, Division	50（插枝） 50 (Cutting)
3月至4月、9月至10月 Mar - Apr; Sep - Oct	播種、插枝、分株 Sowing, Cutting, Division	60（插枝）、100（種子） 60(Cutting), 100 (Seed)
9月至11月 Sep - Nov	播種、分株 Sowing, Division	90-100
9月至11月 Sep - Nov	播種 Sowing	90-100
9月至11月 Sep - Nov	播種、插枝 Sowing, Cutting	100-120
9月至11月 Sep - Nov	播種 Sowing	80-100
2月至5月 Feb - May	播種、插枝 Sowing, Cutting	70-80
3月至5月、9月至10月 Mar - May; Sep - Oct	分株 Division	100-120
2月至5月、9月至11月 Feb - May; Sep - Nov	播種 Sowing	70- 90
3月至5月、9月至11月 Mar - May; Sep - Nov	插枝 Cutting	60
3月至5月 Mar - May	播種、插枝 Sowing, Cutting	70-80
9月至11月 Sep - Nov	播種 Sowing	80-100
10月至12月 Oct - Dec	播種 Sowing	70-90
10月至12月 Oct - Dec	播種 Sowing	90-120
春播、秋播 Spring; Autumn	播種、插枝 Sowing, Cutting	多年生 Perennial

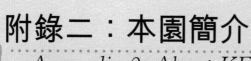

附錄二：本園簡介
Appendix 2: About KFBG

　　嘉道理農場暨植物園(本園)位處新界鄉郊，座落於香港第一高山——大帽山北坡之上，深谷環抱兩旁陡峭的山坡。本園內有清溪、樹林、果園、農圃、遠足徑，並設有動植物展覽、永續農業示範區、藝術展覽、野生動物拯救中心、本土樹木苗圃，以及其他保育及教育設施。

　　本園的歷史可追溯至二戰後的時期。當時，大批貧困的難民湧到香港，其中大部分人擁有傳統農耕和畜牧知識，但缺乏物資；也有些人擁有土地，但沒有農耕經驗。他們都迫切需要援助，以重建新生。有見及此，羅蘭士‧嘉道理和賀理士‧嘉道理於1951年創辦嘉道理農業輔助會，成為香港政府的主要伙伴，協助制定和推行推動香港人自食其力的計劃。嘉道理兄弟出身於商賈世家，卻相信上天賦予他們財富，其實是一種使命，藉此造福人群。在他們的援助下，數以千計的人接受了農耕訓練。此外，輔助會繁殖了數以千計的豬、雞和鴨，贈送或賒售給農夫，亦向數千位農夫提供小型貸款。同時，輔助會興建了許多水井、灌溉水道、道路、步徑、橋樑、豬舍和農舍。1956年，白牛石本來只是個荒蕪的山嶺，輔助會在此開辦農場，作為牲口繁殖及分配、農業研究、農民培訓、公眾教育及康樂的基地。坡地被修築成層層平地，並開墾成果園和菜園。1963年，植物園開始發展，並於1972年開展植物保育計劃。

　　1995年1月20日，香港立法局(現為立法會)通過《嘉道理農場暨植物園公司條例》(香港法例第1156章)。本園正式成為一個致力於保育和教育的非牟利法人團體，並以獨特的公私營合作模式運作，嘉道理農場暨植物園公司雖屬公共機構，但大部分經費來自私營的嘉道理基金(每年超過1億港元)。此外，本園也會收取入場費、課程費用以及市民的捐款，個別項目亦獲得政府資助，這些款項讓本園的工作範圍得以擴展。

　　自1995年起，本園除了在園內舉辦多元化的項目推動自然教育、自然保育及永續生活外，更將其拓展至全港及華南各地。在2015年，我們於大埔開設了綠匯學苑，讓公眾能體驗和學習更多永續生活的方法。

　　現時，全球正經歷幾種危機，包括人與社會、大自然以至內在自我的愈益疏離。人類漠視永續發展的生活模式，不斷耗用及過份依靠地球資源，導致資源損耗速度加快、氣候轉變、生物棲息地萎縮，以及物種消失等問題。作為自然保育機構，本園致力推廣保育意識，同時以精確的科學方法保育物種和修復生態系統，並提倡以新的思維和生活方式應對各種世界問題。因此，我們專注於自然保育、永續生活和整全教育，以促進人類與大自然重新連繫，從而帶來希望和進步。只要大眾、政府、學術界、非政府機構和商界和衷共濟，我們共同的未來才得以守護。

Kadoorie Farm and Botanic Garden (KFBG) is situated in the rural New Territories, on the northern slopes of Tai Mo Shan, Hong Kong's highest mountain. Two steep spurs enclose its deep-set valley. Within KFBG are streams, woodlands, orchards, vegetable gardens, walking trails, live animal exhibits, floral exhibits, sustainable agriculture demonstration plots, art exhibits, a wild animal rescue centre, a native tree nursery, and, other conservation and education facilities.

In the post-war years, Hong Kong was flooded with destitute refugees. Many had traditional knowledge of crop production and livestock farming but no stock, others had land but no experience. They required support to rebuild their lives. In 1951, in response to these pressing human needs Lawrence and Horace Kadoorie established the Kadoorie Agricultural Aid Association (KAAA), which became a key partner of the Hong Kong Government in devising and implementing a plan to help Hong Kong feed itself. The Kadoorie brothers, part of a well-established business family, saw wealth as a sacred trust to benefit mankind. With such aid, thousands of people received agricultural training; thousands of pigs, chickens and ducks were bred and given to farmers or sold to them on credit; thousands received micro-loans; and numerous wells, irrigation channels, roads, footpaths, bridges, pigsties and farm houses were built. The farm site at Pak Ngau Shek was established in 1956 as a base for livestock breeding and distribution, agricultural research, farmer training, public education and recreation. The barren slopes were terraced and planted with orchards and vegetable gardens. The development of the botanic garden began in 1963 and the plant conservation programme in 1972.

On 20th January, 1995, the Legislative Council of Hong Kong passed an Ordinance (KFBG Chapter 1156) incorporating KFBG as a non-profit corporation designated as a conservation and education centre. It is a unique public-private partnership, for while the KFBG Corporation is a public organisation, it is privately funded by the Kadoorie Foundation (over HKD 100 million per year); these funds are supplemented by entrance fees, course fees and small donations from the public and occasional project-related Government grants that enable us to extend our work.

Since 1995, KFBG has been conducting a wide range of nature education, nature conservation and sustainable living programmes both on-site, and, throughout Hong Kong and South China. In 2015, we opened our new, urban facility in Tai Po, the Green Hub, where the public can experience and learn about more sustainable ways of living.

In a time of severe global crisis - including the inter-related issues of widespread disconnection from nature, each other and self; the ever-increasing exploitation of, and unwise over-reliance on the world's dwindling resources to support unsustainable lifestyles; climate change; over population ; shrinking of natural habitats and species loss - KFBG, as an organisation, raises awareness, undertakes rigorous science-based species conservation and ecosystem restoration, and offers new ways of thinking and living to respond to the world's problems. Hence, our work brings hope and improvement by focusing on nature conservation, sustainable living and holistic education that re-connects people with nature. By working together with the public, Governments, academia, NGOs and businesses, we can protect our common future.

參考書目
Reference

❖ Bill Mollison, 2002, "Permaculture: A Designers' Manual", Tagari Publications, Australia.

❖ Bill Mollison, 2009, "Introduction to Permaculture", Tagari Publications, Australia.

❖ Basil Caplan, 1992, "Organic Gardening", Headline Book Publishing, London.

❖ Michel & Jude Fanton, 1993, "The seed savers' handbook", Seed Savers' Network, Australia.

❖ Richard Bird, 2000, "The fruit & Vegetable Garden", Lorenz Books, London.

❖ Mel Bartholomew, 1981, "Square Foot Gardening" Rodale Press, Emmaus, Pennsylvania.

❖ Graham Bell, 1994, "The Permaculture Garden", Thorsons, London.

❖ 康有德等，2002年，《園藝概論》，啓英文化事業有限公司，台北。

❖ 趙漢珪，2004年，《自然農業》，延邊大學出版社，吉林。

❖ 彼得·羅賓森，2001年，《省水花園》，貓頭鷹出版社股份有限公司，台北。

❖ 薛聰賢，2000年，《台灣蔬果實用百科》1、2冊，台灣普綠文化事業有限公司，台灣。

❖ 林登·霍桑·黛妮·波恩，2002年，《芳香藥草園藝圖鑑》，貓頭鷹出版社股份有限公司，台北。

❖ 宮野弘司等，2003年，《香草栽培事典》，晨星出版有限公司，台中。

❖ Geoff Hamilton，1999年，《有機種植完全手冊》，貓頭鷹出版社股份有限公司，台北。

❖ 中華人民共和國北京植物檢疫局，1999年，《中國植物檢疫性害蟲圖冊》，中國農業出版社，北京。

❖ 王洪久、曲存英，2002年，《蔬菜病蟲害原色圖譜》，山東科學技術出版社，濟南。

❖ 商鴻生、王鳳葵、張敬澤，2003年，《綠葉菜類蔬菜病蟲害診斷與防治原色圖譜》，金盾出版社，北京。

❖ 商鴻生、王鳳葵、徐秉良，2003年，《白菜甘藍蘿蔔類蔬菜病蟲害診斷與防治原色圖譜》，金盾出版社，北京。

❖ 願耘、李桂舫、張迎春，2003年，《豆類蔬菜病蟲害診斷與防治原色圖譜》，金盾出版社，北京。

❖ 楊秋忠，1999年，《土壤與肥料》，農世股份有限公司，台中。

❖ 2002年，《陽台蔬菜園藝》，科學技術文獻出版社，北京。

❖ 劉保才，1999年，《蔬菜高產栽培技術大全》，中國林業出版社，北京。

❖ 2001年，《家庭蔬果盆栽》，瑞昇文化事業股份有限公司，台北。

❖ 張可欣等，2002年，《完全校園有機耕種手冊》，綠田園基金，香港。

❖ 2012年，《香港植物名錄》，漁農自然護理署，香港。

❖ 2005年，《從三斤半菜開始》，嘉道理農場暨植物園，香港。

參考網頁

❖ 嘉道理農場暨植物園（http://www.kfbg.org/eng/）
Kadoorie Farm and Botanic Garden

❖ 綠匯學苑（www.greenhub.hk）
Green Hub

❖ 漁農自然護理署（http://www.afcd.gov.hk/cindex.html）
Agriculture, Fisheries and Conservation Department

❖ 香港植物標本室（http://herbarium.gov.hk/）
Hong Kong Herbarium

❖ 香港有機資源中心認證有限公司（http://www.hkorc-cert.org/tc/）
Hong Kong Organic Resource Centre Certification Ltd.

❖ The Permaculture Research Institute（http://www.permaculture.org.au/resources/）